Convolution Type Functional Equations

———————— on ————————

Topological
Abelian
Groups

Convolution Type Functional Equations

on

Topological Abelian Groups

László Székelyhidi

Mathematical Institute
University of Debrecen
Hungary

World Scientific
Singapore • New Jersey • London • Hong Kong

Published by

World Scientific Publishing Co. Pte. Ltd.

P O Box 128, Farrer Road, Singapore 9128

USA office: 687 Hartwell Street, Teaneck, NJ 07666

UK office: 73 Lynton Mead, Totteridge, London N20 8DH

Library of Congress Cataloging-in-Publication data is available.

CONVOLUTION TYPE FUNCTIONAL EQUATIONS ON TOPOLOGICAL ABELIAN GROUPS

ISBN 981-02-0658-5

Printed in Singapore by JBW Printers & Binders Pte. Ltd.

To my family and friends

To my family and friends

PREFACE

In this volume we make an attempt to present a unified treatment for a branch of the theory of functional equations: convolution type functional equations on topological abelian groups. The main idea is to use the fundamental results about spectral synthesis on discrete abelian groups. The content of these results is, roughly speaking, that a translation invariant linear space of complex valued functions on a discrete abelian group, which is closed with respect to pointwise convergence, is determined by the exponential monomials included in it. More exactly, the linear hull of the set of all exponential monomials contained in a translation invariant closed linear space of complex valued functions on a discrete abelian group is everywhere dense in this linear space. Here "closed" means "closed under formation of pointwise limits". This is the discrete version of the famous result of L.Schwartz [SCZ1] on spectral synthesis on the real line, which has been generalized for special varieties on locally compact abelian groups (see Section 8). From the point of view of functional equations this result means, that if a functional equation (or a system of functional equations) has the property, that all complex valued solutions of it form a translation invariant closed linear space, then any solution is a pointwise limit of exponential polynomial solutions. Hence in order to characterize all solutions one needs to determine only the exponential polynomial, or rather, the exponential monomial solutions. This approach provides a unified treatment for a large class of functional equations; roughly speaking, for all functional equations of the form

$$f * \mu = 0,$$

where μ is any nonzero compactly supported Radon measure. Obviously, systems of such equations can also be considered. We note that one actually needs the spectral synthesis on finitely generated discrete abelian groups only, as it will be seen in the sequel.

The present volume is devoted to some classical functional equations. These are: functional equations characterizing generalized polynomials, Levi-Civitá-type functional equations, d'Alembert-type functional equations, difference equations, and mean-value type functional equations. It will be shown, that all these functional equations (being special convolution type equations) can be studied on the basis of spectral synthesis. Further, we derive also some technical consequences concerning ordinary and partial difference equations.

This volume consists of 17 sections. The first section is an introduction to the basic material of spectral analysis and synthesis which is needed in the sequel and a motivation of the forthcoming results. The following 6 sections are devoted to the study of exponential polynomials on topological abelian groups.

Section 1 is an introduction, containing well-known results and terminology about multi-additive functions on semigroups.

In Section 2 polynomials on semigroups are introduced. The results are useful to invert representation theorems into characterization theorems, concerning polynomials on semigroups and groups.

The theorems included in Section 3 show that mild regularity properties of polynomials imply their strong regularity: continuity. This is a classical field: the strong analytical consequences of mild regularity conditions of polynomial functions have attracted the attention of several authors.

Section 4 is devoted to the introduction of exponential polynomials on semigroups, which play an utmost important role in the investigations. The slight distinction between normal exponential polynomials and exponential polynomials is important from the point of view of spectral synthesis.

Section 5 is an analogue of Section 3 in the sense, that here the pleasant analytical behavior of exponential polynomials is presented.

Section 6 is a copy of the appropriate sections of the "bible" of functional equations: the book of J.Aczél [ACZ7]. Here the particular form of exponential polynomials on some special groups and semigroups is given.

In Section 7 a useful tool: the Fourier-transform of exponential polynomials is introduced. This is an analogue of the Fourier-transform of almost periodic functions, and it can be used successfully to determine exponential polynomial solutions of functional equations.

Section 8 deals with mean periodic functions and spectral synthesis on discrete abelian groups.

Section 9 is devoted to functional equations characterizing polynomials. Most equations and results here are well-known and have been taken from [ACZ7], but here the proofs are based on spectral synthesis.

Section 10 includes a complete description of the nondegenerate solutions of the Levi-Cività equation, based on the methods assured by spectral synthesis. Special applications for well-known functional equations are also given.

Section 11 is devoted to d'Alembert-type functional equations. The problem on the general solution of this type of equations is completely solved on the basis of spectral synthesis. We note that no divisibility conditions on the underlying group structure will be posed.

In Section 12 special applications of the results of Section 10 and Section 11 are given: addition and subtraction theorems. Among others the classical functional equations describing the addition theorems of the trigonometric polynomials are re-considered.

Section 13 is devoted to difference equations. New results are presented concerning general difference equations on compactly generated locally compact abelian groups.

Section 14 contains applications concerning functions of several variables. Functional equations of mean-value type have been treated by different authors. Here we present the solution of two problems concerning the octahedron and the cube equation. First we show (proving the conjecture of D.Z.Djokovič and H.Haruki) that the generalized octahedron and cube equations are equivalent. In the regular case it is well-known that the solutions are special harmonic polynomials. We prove (according to another conjecture of H.Haruki) that all regular solutions are linear combinations of partial derivatives of a given harmonic polynomial.

Section 15 is devoted to special applications concerning ordinary and partial differential equations. Here the Fourier-transform of exponential polynomials has been used to produce explicit solutions of ordinary differential equations and Cauchy-problems for basic partial differential equations, if the given data are exponential polynomials.

Although the methods in this volume depend heavily on commutativity, there is some place for noncommutative applications. These are summarized in Section 16 by presenting some examples.

In preparing this work we have used constantly the comprehensive volume [ACZ7]. Concerning measure theory we use the results and terminology of [FED] and [HAL]. The basic material on spectral theory has been taken from [LOO]. The reference list includes those items which had the most effective influence on our work.

This work has been completed during my stay at the University of Hamburg as a research fellow of the Alexander von Humboldt Foundation. I am indebted to the Foundation, to the Mathematisches Seminar der Universität Hamburg and to Prof. Walter Benz for their valuable help.

Debrecen, 1990   László Székelyhidi

CONTENTS

PRELIMINARIES

THE BASIC PROBLEMS OF SPECTRAL ANALYSIS AND SPECTRAL SYNTHESIS

A basic tool for the investigation of different algebraic or analytical structures is duality. Duality makes it possible to represent abstract structures by more special ones; for instance abstract sets, topological spaces, linear spaces, normed spaces, algebras, topological groups, etc., by similar structures of functions which are defined on some particular sets. In the theory of algebras the corresponding representation process can be described by the *Gelfand-transformation*.

Let A be an algebra and let H denote a set of *algebra homomorphisms* of A onto $\mathbf{C}$. Such homomorphisms are called *multiplicative linear functionals*.

The Gelfand-transformation on A is defined by

$$\hat{x}(h) = h(x)$$

for any x in A and h in H. Then $\hat{x} : H \to \mathbf{C}$ is a function and obviously $x \to \hat{x}$ is an algebra homomorphism of A onto the function algebra of all functions of the form $\hat{x}$, which we denote by $\hat{A}$. The function $\hat{x}$ is called the *Gelfand-transform* of x.

In order to "represent" A as the function algebra $\hat{A}$, it is necessary to study the following basic problems:

i) When is $x \mapsto \hat{x}$ one-to-one?
ii) An internal characterization (or identification) of H is necessary.
iii) A characterization of $\hat{A}$ is necessary: which functions on H belong to $\hat{A}$? In particular, when do "all" functions on H belong to $\hat{A}$?

For the first problem, it is easy to see, that $x \mapsto \hat{x}$ is one-to-one if and only if H is a *separating family* for A, that is, for all $x \neq y$ there exists an h in H with $h(x) \neq h(y)$. As H consists of homomorphisms, it is enough to have for any nonzero x in A an h with $h(x) \neq 0$. This means, that the nonzero elements of the intersection of the kernels of all elements of H will violate

1

injectivity. As obviously any element of the form $xy - yx$ belongs to this intersection, for the injectivity of the Gelfand-transformation it is necessary, that A is *commutative*. This is a quite natural assumption, for in order to represent A as a function algebra it must be commutative, since function algebras are. However, commutativity is not sufficient for the Gelfand-transformation is one-to-one.

The next step is to identify H. It is obvious, that for any h in H the kernel of h is a *maximal ideal*. Maximality follows from the *homomorphism theorem*: $A/Ker(h)$ is isomorphic to $\mathbf{C}$, which has no proper ideals, hence there are no intermediate ideals between $Ker(h)$ and A. Another property of $Ker(h)$ is *regularity*: there exists an element u which is a *relative identity* with respect to $Ker(h)$: $ux \equiv x \mod Ker(h)$, that is, $ux - x \in Ker(h)$ for all x. Indeed, any u with $h(u) = 1$ satisfies this property. We note, that in general, an ideal is called *regular*, if there exists such a relative identity with respect to it. Of course, in a commutative algebra with identity any ideal is regular.

Hence, to any h in H there corresponds a regular maximal ideal M_h which is the kernel of h. It is easy to see, that this correspondence is one-to-one, that is, different multiplicative linear functionals cannot have the same kernel. For a bijective correspondence between multiplicative linear functionals and regular maximal ideals one needs a statement like this: any regular maximal ideal is the kernel of some multiplicative linear functional. In any case the regular maximal ideal M is the kernel of the *natural homomorphism* $A \to A/M$, and by the maximality of M, A/M is a field. The question is: in which case can A/M be identified with $\mathbf{C}$? The following theorem helps:

THEOREM OF GELFAND-MAZUR. *Any normed field is isometrically isomorphic to the complex field.*

Hence it is useful to assume, that A is a normed algebra. But this is not enough in general to assure, that A/M is a normed field. For, the standard technique, taking the norm from A onto A/M, works only if M is closed. The problem is: in which case will any regular maximal ideal of A be closed? To answer this question, we need the notion of *adverse*. The adverse of an element x is the element y, if $xy - x - y = 0$. Heuristically, $y = \frac{x}{x-1}$, but there is no identity in general. The adverse is just to play the role of an inverse in the lack of an identity. Now the following theorem holds:

THEOREM. *In a commutative normed algebra every regular maximal ideal is closed if and only if every element x with $\| x \| < 1$ has an adverse.*

If the series converges, then $y = -\sum_1^\infty x^k$ is an adverse of x. On the other hand, if $\parallel x \parallel < 1$, then this series is *absolutely convergent*. What we need is, that any absolutely convergent series is convergent. But this is the case if and only if A is a *Banach-space*. This is the natural setting for the theory of Gelfand-transformation.

Suppose, that H consists of all continuous multiplicative linear functionals of A. Now the correspondence $h \rightarrow Ker(h)$ is bijective, since every regular maximal ideal is closed, thus it is the kernel of a continuous multiplicative linear functional. Hence this correspondence identifies H with the space of all regular maximal ideals of A: with the *maximal ideal space* of A, which is denoted by Δ.

Returning to the Gelfand-transformation, for its bijectivity it is necessary and sufficient, that Δ is separating for A. In the language of functionals this means that there are enough continuous multiplicative functionals in order to separate the elements of A. In the language of maximal ideals this means, that there exists no nonzero x which belongs to all regular maximal ideals, that is, the intersection of all regular maximal ideals is zero. In general this intersection is called the *radical* of the algebra, and if it is zero, then A is called *semi-simple*. Of course, in general it is a nontrivial problem to decide if a given commutative Banach-algebra is semi-simple.

Usually Δ is given the *weak topology* induced by $\hat{A}$, that is, the weakest topology with respect to which all Gelfand-transforms are continuous. This is the same, as the weak $\star$ topology for Δ, if it is considered as a subset of A^*, the dual of A as a Banach-space. It is quite easy to see, that Δ is a subset of the unit ball in A^*, hence its closure is compact in the weak $\star$ topology, by the theorem of Banach-Alaoglu (see e.g. [HEW], [LOO]). If A has an identity, then Δ is closed itself, hence it is compact. In general, Δ is locally compact and Hausdorff. If Δ is not compact, then all functions in $\hat{A}$ vanish at infinity.

Summarizing these results, if A is a commutative Banach-algebra and Δ is the (locally compact Hausdorff) space of all regular maximal ideals of A, then the Gelfand-transformation $x \mapsto \hat{x}$ is a norm-decreasing algebra homomorphism of A into the function algebra $\hat{A}$, which consists of continuous functions vanishing at infinity and defined on Δ. The central problems of the theory of commutative Banach-algebras concern with finding conditions to assure, that this Gelfand-transformation is one-to-one (the algebra is semi-simple), and is onto the space $\mathcal{C}_0(\Delta)$. If it is, then the algebra can be considered as the algebra of all continuous complex valued functions, vanishing at infinity, on a locally compact Hausdorff space.

A very simple, but important example is the following. Let X be a com-

3

pact Hausdorff space and let $A = \mathcal{C}(X)$ be the set of all continuous complex valued functions on X, which is a commutative Banach-algebra with identity, if it is equipped with the sup-norm. For any $p \in X$ the set

$$M_p = \{f : f(p) = 0\}$$

is a closed ideal in $\mathcal{C}(X)$, and the maximal ideals are exactly the ones of the form M_p with $p \in X$. The correspondence $p \mapsto M_p$ is bijective, and hence, set-theoretically, Δ can be identified with X. For this correspondence is a homeomorphism, it is necessary and sufficient, that the topology of X is identical with the one, defined by $\mathcal{C}(X)$, but this is just the *complete regularity* of X, that is the topology of X is completely determined by the continuous functions. This is a consequence of compactness.

The Gelfand-transform for $f \in \mathcal{C}(X)$ is

$$\hat{f}(M_p) = h_p(f) = f(p)$$

that is, as Δ is identified with X, $\hat{f}$ can be identified with f.

Similar arguments can be used if X is a locally compact Hausdorff space and $A = \mathcal{C}_0(X)$ is the algebra of all continuous complex valued functions on X, vanishing at infinity.

Hence, in the case of the ideals of $\mathcal{C}(X)$ any proper ideal is included in a maximal ideal. An application of Zorn's lemma gives the same for any proper regular ideal in any commutative Banach-algebra.

THEOREM. *In a commutative Banach-algebra any proper regular ideal is included in a regular maximal ideal.*

Now one can formulate the basic problem of spectral analysis: Is every proper closed ideal of a commutative Banach-algebra included in a regular maximal ideal? The answer is "yes" in $\mathcal{C}(X)$, but "no" in general. A basic theorem in this context is

THE WIENER TAUBERIAN THEOREM. *Let A be a regular semi-simple Banach algebra with the property that the set of those elements x for which $\hat{x}$ has compact support is dense in A. Then every proper closed ideal is included in a regular maximal ideal.*

Here "regular" means, that the algebra is commutative, and for any closed set $C \subset \Delta$ and for any $M \notin C$ there exists a function $\hat{f} \in \hat{A}$ with $\hat{f}(M) \neq 0$ and $\hat{f} = 0$ on C.

4

The formulation of the basic problem of spectral synthesis is the following: Is every proper closed ideal of a commutative Banach-algebra the intersection of all regular maximal ideals, in which it is contained? The answer is again "yes" for $\mathcal{C}(X)$, where X is compact Hausdorff, and "no" for the general case.

Now we specialize our setting by restricting ourselves to the algebra $L^1(G)$, where G is a *locally compact abelian group*, and multiplication is defined by *convolution*. With the L^1-norm this is a commutative Banach-algebra, which has an identity if and only if G is *discrete*. For the characterization of the maximal ideal space of $L^1(G)$ one knows, that for any M in Δ there exists a multiplicative linear functional on $L^1(G)$ with the kernel M. As this functional is continuous and linear on $L^1(G)$, and the dual of $L^1(G)$ can be identified with $L^\infty(G)$, this functional can be uniquely represented by an element α_M of $L^\infty(G)$ in the form

$$\hat{f}(M) = \int f(x)\overline{\alpha_M}(x)dx.$$

From the multiplicativity of α_M it follows that

$$\alpha_M(x+y) = \alpha_M(x)\alpha_M(y)$$

holds almost everywhere on $G \times G$. Then it follows by standard methods that α_M is almost everywhere equal to a continuous function with absolute value 1 satisfying the above functional equation for all x and y in G. This means, that α_M can be identified with a function, which we call a *character* of G. Thus the maximal ideal space of $L^1(G)$ can be identified with the set $\hat{G}$ of all characters of G. Through this identification it follows that $\hat{G}$ is locally compact in the weak $\star$ topology of $L^\infty(G) = L^1(G)^*$. And it is easy to see, that $\hat{G}$ is a locally compact abelian group, multiplication being defined as the pointwise multiplication of functions. $\hat{G}$ is called the *dual* of G. It follows, that $\hat{G}$ is compact if and only if G is discrete. Further, the Gelfand-transformation has the following form: for $f \in L^1(G)$ we have

$$\hat{f}(M) = \int f(x)\overline{\alpha_M}(x)dx,$$

where α_M is the character corresponding to the regular maximal ideal M. The function $\hat{f} : \hat{G} \to \mathbf{C}$ is called the *Fourier-transform* of the function f.

We summarize the basic facts on the Fourier-transformation. For any locally compact abelian group G the Fourier-transformation $f \mapsto \hat{f}$ is a norm-decreasing algebra homomorphism of $L^1(G)$ onto a set of continuous, at infinity vanishing complex valued functions, defined on the (locally compact abelian)

5

dual group $\hat{G}$ of G. (The statement, that the Fourier-transforms of the functions in $L^1(G)$ are vanishing at infinity is called Mercer's theorem in the case $G = \mathbf{T}$, and Riemann-Lebesgue-lemma in the case $G = \mathbf{R}$.)

Without going into the details of the duality theory of locally compact abelian groups let's state here some consequences of the famous Pontryagin's duality theorem:

i) The dual of the dual of G is G.
ii) The dual of a discrete group is compact; the dual of a compact group is discrete.
iii) The dual of $\mathbf{Z}$ is $\mathbf{T}$, the dual of $\mathbf{T}$ is $\mathbf{Z}$, the dual of $\mathbf{R}$ is $\mathbf{R}$.

We collect here also some basic theorems on the Fourier-transformation. Recall, that a function p in $L^\infty(G)$ is called *positive definite*, if the respective linear functional on $L^1(G)$ is positive, that is, if

$$\int \int f(x)\overline{f(y)}p(x - y)dxdy \geq 0$$

for all f in $L^1(G)$. For instance, any character is positive definite. It follows, that convex combinations of positive definite functions are positive definite again. The famous theorem of Bochner states, that if we allow "continuous" convex combinations, that is, integrals with respect to positive measures, then all positive definite functions have this form.

THEOREM OF BOCHNER. *The formula*

$$p(x) = \int \alpha(x)d\mu(\alpha)$$

sets up a norm preserving isomorphism between the convex set of all finite positive Baire-measures μ on $\hat{G}$ and the convex set of all positive definite functions p in $L^\infty(G)$.

The following theorem is on the inverse of the Fourier-transformation restricted to a special set concerning positive definite functions. We denote by $[L^1(G) \cap P]$ the closed subspace generated by the positive definite functions in $L^1(G)$.

INVERSION THEOREM. *If f is in $[L^1(G) \cap P]$, then $\hat{f}$ is in $L^1(\hat{G})$ and*

$$f(x) = \int \hat{f}(\alpha)\alpha(x)d\alpha$$

for almost all x in G, where $d\alpha$ is the Haar-measure on $\hat{G}$, suitably normalized.

One more important theorem in this relation is the

THEOREM OF PLANCHEREL. *The Fourier-transformation on $[L^1(G) \cap P]$ preserves scalar product and its L^2-closure is a unitary operator of $L^2(G)$ onto $L^2(\hat{G})$.*

Returning to spectral analysis and spectral synthesis, we have

THEOREM. $L^1(G)$ *is semi-simple and regular.*

This means, that if the condition, that the elements, whose Fourier-transforms vanish off compact sets are dense in $L^1(G)$ is satisfied, then one can apply the Wiener Tauberian Theorem for $L^1(G)$. But it is easy to check, that this is the case, hence the Tauberian Theorem for $L^1(G)$ can be restated.

TAUBERIAN THEOREM. *If G is a locally compact abelian group, then every proper closed ideal of $L^1(G)$ is included in a maximal ideal.*

This means, that spectral analysis holds for the proper closed ideals of $L^1(G)$.

This theorem implies the following approximation theorem of Wiener:

APPROXIMATION THEOREM OF WIENER. *If the Fourier-transform of $f \in L^1(G)$ never vanishes, then the translates of f generate $L^1(G)$.*

This theorem depends on the fact that the closed ideals of $L^1(G)$ are just its closed translation invariant subspaces. Having this in mind, by the Tauberian Theorem f belongs to no regular maximal ideal. Hence the closed ideal generated by f, that is the closed subspace generated by the translates of f is the whole of $L^1(G)$. This is a typical spectral analysis theorem: if the translates of f do not generate $L^1(G)$, then the Fourier-transform of f vanishes somewhere.

As $L^\infty(G)$ is the dual of $L^1(G)$, several problems on $L^1(G)$ can be formulated in the dual language, as problems in $L^\infty(G)$. We note, that L^∞ is equipped with the weak $\star$ topology, and all topological notions in $L^\infty(G)$ relate

7

to this topology. From the general dual space theory we know the notion of the *annihilator*. If $I \subset L^1(G)$ is a subspace, then let

$$I^\perp = \{f \in L^\infty(G) : \langle f, \varphi \rangle = 0 \ for \ all \ \varphi \ in \ I\},$$

and if $V \subset L^\infty(G)$ is a subspace, then let

$$V^\perp = \{\varphi \in L^1(G) : \langle f, \varphi \rangle = 0 \ for \ all \ f \ in \ V\}.$$

Then $I^\perp, V^\perp$ are closed subspaces in $L^\infty(G)$ and $L^1(G)$, respectively, and from the Banach-space theory one knows, that $(I^\perp)^\perp = I$ for any closed subspace I in $L^1(G)$. As the dual of $L^\infty(G)$ with respect to the weak $\star$ topology is just $L^1(G)$, we have also $(V^\perp)^\perp = V$ for any closed subspace V in $L^\infty(G)$. Now there are two one-to-one correspondences between the closed subspaces of $L^1(G)$ and $L^\infty(G)$, which are inverse to each other. By the Hahn-Banach theorem it follows, that at this correspondence ideals of $L^1(G)$ correspond to invariant (that means, translation invariant) subspaces of $L^\infty(G)$.

THEOREM. *Let $V \subset L^\infty(G)$ be a closed subspace. Then $V^\perp$ is a closed ideal in $L^1(G)$ if and only if V is translation invariant.*

Hence there is a one-to-one correspondence between the closed ideals of $L^1(G)$ and the closed invariant subspaces of $L^\infty(G)$. The characterization of the closed ideals of $L^1(G)$ is immediate.

THEOREM. *A closed subset $I \subset L^1(G)$ is an ideal if and only if it is an invariant subspace.*

Now we translate our basic theorems on spectral analysis and spectral synthesis into the language of $L^\infty(G)$. What does it mean for $I^\perp$, that I is contained in a regular maximal ideal in $L^1(G)$? Or simply: what does it mean for $I^\perp$, that I is a regular maximal ideal in $L^1(G)$? Then $I^\perp$ has no proper closed invariant subspace, hence it is a minimal proper closed invariant subspace. On the other hand, as I is the kernel of some multiplicative linear functional, that is, $I = \{\varphi : \hat\varphi(\gamma) = 0\}$ for some $\gamma \in \hat{G}$, it follows, that $\gamma \in I^\perp$ and then $I^\perp = \tau(\gamma) = \{c\gamma : c \in \mathbf{C}\}$. That is, closed proper minimal invariant subspaces are just the one dimensional subspaces, generated by characters. In other words, $I \subset L^1(G)$ is a regular maximal ideal if and only if $I^\perp$ is a one dimensional subspace generated by a character.

8

What does spectral analysis for a proper closed ideal $I \subset L^1(G)$ imply for $I^\perp$? The Wiener Tauberian Theorem can be reformulated.

THEOREM. *Any nonzero closed invariant subspace of $L^\infty(G)$ contains a character.*

This is the first point, where an example can be given for the possible applications of the results of spectral analysis in the theory of functional equations. Namely, one can solve the classical *d'Alembert equation* in $L^\infty(G)$.

THEOREM. *Let G be a locally compact abelian group and $f : G \to \mathbf{C}$ a nonzero bounded and measurable function satisfying*

$$f(x + y) + f(x - y) = 2f(x)f(y)$$

for all x and for almost all y in G. Then

$$f(y) = \frac{1}{2}(\gamma(y) + \gamma(-y))$$

holds for almost all y in G with some character γ of G.

PROOF: As f is nonzero, $\tau(f) \subset L^\infty(G)$ contains a character γ. But any g in $\tau(f)$ satisfies

$$g(x + y) + g(x - y) = 2g(x)f(y)$$

for all x and for almost all y in G, hence we have for γ

$$\gamma(x + y) + \gamma(x - y) = 2\gamma(x)f(y)$$

for all x and for almost all y in G. Dividing by $\gamma(x) \neq 0$ we get our statement:

$$f(y) = \frac{1}{2}(\gamma(y) + \gamma(-y))$$

for almost all y in G.

The main problem of spectral analysis for $L^1(G)$ was: is every proper closed ideal of $L^1(G)$ contained in a regular maximal ideal? This has been reformulated for $L^\infty(G)$: does every proper closed invariant subspace of $L^\infty(G)$ contain a character? The answer is "yes" in both cases, due to the Wiener

9

Tauberian Theorem. Spectral analysis of a proper closed invariant subspace of $L^\infty(G)$ means to determine the characters in this subspace. If one takes $f \in L^\infty(G)$, then spectral analysis of f means to determine the characters of $\tau(f)$. These are the candidates for taking part in the spectral synthesis of f, that is, in the reconstruction process of f from characters. The set of all characters in $\tau(f)$ (or more general, in any proper closed invariant subspace $V \subset L^\infty(G)$) is called the *spectral set* of f (the spectral set of V, respectively). Notation: $sp(f)$ or $sp(V)$. It is easy to see, that the spectral set of f is identical with the set of common zeros of the Fourier-transforms of the φ's in $L^1(G)$, annihilating $\tau(f)$. That is, γ belongs to the spectral set of f if and only if $\hat{\varphi}(\gamma) = 0$ for all φ in $L^1(G)$ with $f * \varphi = 0$.

Let us, for instance, determine the spectral set of a character γ in $L^\infty(G)$. As $\tau(\gamma)$ consists of the scalar multiples of γ, the only character in $\tau(\gamma)$ is γ. Hence the spectral set of any character γ is $\{\gamma\}$. We may use also the other approach to determine $sp(\gamma)$. Namely, χ is in $sp(\gamma)$ if and only if $\hat{\varphi}(\chi) = 0$ for all φ in $L^1(G)$ with $\gamma * \varphi = 0$, that is, with $\hat{\varphi}(\gamma) = 0$. It means $\chi = \gamma$. Similarly, if $\sum_{i=1}^{n} c_i \gamma_i$ is a *trigonometric polynomial*, then its spectral set is $\{\gamma_1, \gamma_2, \ldots, \gamma_n\}$. We repeat it again: the elements of the spectral set of f in $L^\infty(G)$ are those characters, which are weak $\star$ limits of nets formed by linear combinations of translates of f. Practically these can be computed, if we determine all common zeros of the Fourier-transforms of the elements in $\tau(f)^\perp$, that is, of the functions φ in $L^1(G)$, with $f * \varphi = 0$. Suppose, for instance, that f is in $L^1(G) \cap L^\infty(G)$. Then $f * \varphi = 0$ implies $\hat{f}\hat{\varphi} = 0$, hence the spectral set of f is identical with the support of $\hat{f}$. This is quite natural, if we think of the inversion theorem.

By the definition of the spectral set of f in $L^\infty(G)$ it follows, that if $f * \varphi = 0$ for some φ in $L^1(G)$, then $\hat{\varphi} = 0$ on $sp(f)$. The converse holds under a stronger assumption.

THEOREM. *If $\varphi \in L^1(G)$ vanishes on a neighborhood of $sp(f)$ for some $f \in L^\infty(G)$, then $f * \varphi = 0$.*

The question, whether $f * \varphi = 0$ on the hypothesis merely that φ vanishes on $sp(f)$, was the celebrated problem of A.Beurling ([BEU], [HEL2]).

THE BEURLING PROBLEM. *If $\varphi \in L^1(G)$ vanishes on $sp(f)$ for some $f \in L^\infty(G)$, does it follow $f * \varphi = 0$?*

It can be shown, that the Beurling problem has an affirmative answer if

10

and only if f is the weak $\star$ limit of trigonometric polynomials in $sp(f)$, or equivalently, if for $\tau(f)^\perp$ in $L^1(G)$ spectral synthesis holds.

Now we can summarize, what does it mean for $I^\perp$ in $L^\infty(G)$, that spectral synthesis holds for I in $L^1(G)$: this means precisely, that the trigonometric polynomials of $I^\perp$ are (weak $\star$) dense in $I^\perp$. A negative solution for the Beurling problem in $\mathbf{R}^3$ was given by L.Schwartz in 1947 ([SCZ2]), who presented a counterexample. From a previous theorem we know, that any element of $\tau(f)$ can be approximated by trigonometric polynomials, taken from a neighborhood of $sp(f)$. It follows, that if $\hat{G}$ is discrete, then we have a positive solution for the Beurling problem. This is the half of the following theorem ([MAL]).

THEOREM OF MALLIAVIN. *Spectral synthesis holds for $L^1(G)$ if and only if G is compact.*

This means, that all proper closed ideals have spectral synthesis if and only if G is compact. Nevertheless, in general there may exist some special proper closed ideals in $L^1(G)$, for which spectral synthesis holds. The following theorem presents a case of this type.

PRIMARY IDEAL THEOREM. *If the spectral set of a bounded measurable function has exactly one point, then the function is a constant multiple of a character.*

This theorem for any locally compact abelian group is due to I. Kaplansky ([KAP]), who used the structure theory of locally compact abelian groups for his proof. Previously the theorem has been proved for $G = \mathbf{R}$ by V.Ditkin ([DIT]). An independent proof based on distribution theory was given by J.Riss ([RIS]). Another proof for the general case which does not depend on the structure theory was given by H.Helson ([HEL1]). The theorem has its name, as the dual statement reads: a closed ideal of functions in $L^1(G)$, whose Fourier-transforms have only one common zero necessarily contains all functions, whose transforms vanish at that point. Ideals, which are contained in precisely one regular maximal ideal are called *primary ideals*. A simple analogue of the primary ideal theorem in the case of $\mathcal{C}(X)$ is the following: if the functions of a proper closed ideal in $\mathcal{C}(X)$ have only one common zero, then this ideal is just the maximal ideal of all functions, vanishing at that point.

Extensions of the primary ideal theorem on the real line are due to V.Ditkin, I.E.Segal, S.Mandelbrojt and S.Agmon ([DIT], [MGM], [SEG2],). These are of the following type: if the boundary of the spectral set of an $f \in L^\infty(G)$

is denumerable, or it does not contain any nonempty perfect set, then f is a limit of trigonometric polynomials of $sp(f)$.

Our last formulation of the basic spectral problems was the following:

i) Does every proper weak $\star$ closed invariant subspace of $L^{\infty}(G)$ contain a character?

ii) Are the trigonometric polynomials of every proper weak $\star$ closed invariant subspace of $L^{\infty}(G)$ dense in this subspace?

This formulation depends on the fact that $L^{\infty}(G)$ is the dual of $L^1(G)$. But we can consider this problem from a more general point of view. Namely, the dual pair $(L^1(G), L^{\infty}(G))$ can be substituted by several other dual pairs $(F(G), F(G)^*)$, where $F(G)$ is a given *linear translation invariant function space* and $F(G)^*$ is given the weak $\star$ topology. Then the two basic problems have the following form:

i) Does every proper closed invariant subspace of $F(G)$ contain "minimal" invariant subspaces?

ii) Do the minimal invariant subspaces of every proper closed invariant subspace of $F(G)$ "generate" this subspace?

Of course, a precise meaning must be given the words "minimal" and "generate". We have seen, that if a convolution in $F(G)^*$ can be defined, then the "minimal" invariant subspaces of $F(G)$ relate somehow to the multiplicative linear functionals of the algebra $F(G)^*$. In order to understand the situation better we present here the simple example of $G = \mathbf{Z}$, where $F(\mathbf{Z})$ is the set of all complex valued functions on $\mathbf{Z}$. We equip $F(\mathbf{Z})$ with the topology of pointwise convergence, hence $F(\mathbf{Z})^*$ is the set of all finitely supported complex measures on $\mathbf{Z}$, which can also be realized as the set of all finitely supported complex valued functions on $\mathbf{Z}$. Then $F(\mathbf{Z})^*$ is a commutative (non-Banach!) algebra with identity and its maximal ideal space can be identified with $\mathbf{C} - \{0\}$ in the following manner: any maximal ideal of $F(\mathbf{Z})^*$ is the kernel of a multiplicative linear functional h_λ with $\lambda \in \mathbf{C}, \lambda \neq 0$, which has the form

$$h_\lambda(\mu) = \int \lambda^{-n} d\mu(n) = \sum_{k=-N}^{N} c_k \lambda^k,$$

if $\mu = \sum_{k=-N}^{N} c_k \delta_{-k}$, where δ_{-k} is the *Dirac-measure*, concentrated at the point $-k$. Conversely, any multiplicative linear functional h on $F(\mathbf{Z})^*$ has the above form with some nonzero complex λ and the correspondence between h_λ and λ is one-to-one. As $\hat{\mathbf{Z}} = \mathbf{T}$, one can see that the characters of $\mathbf{Z}$ are not sufficient in this situation. But any nonzero complex λ defines a function $m_\lambda(n) = \lambda^n$ on $\mathbf{Z}$ which has the fundamental property of characters

$$m_\lambda(n + k) = m_\lambda(n)m_\lambda(k),$$

although it is not necessarily bounded. It is called a *generalized character* or an *exponential* of $\mathbf{Z}$. Hence the maximal ideal space of $F(\mathbf{Z})^*$ can be identified with the set of all exponentials of $\mathbf{Z}$.

Now we consider any nonzero μ of the form

$$\mu = \sum_{k=0}^{N} c_k \delta_{-k}$$

in $F(\mathbf{Z})^*$. If f is an element of the annihilator of the ideal generated by μ, then f is a solution of the difference equation

$$f * \mu = \sum_{k=0}^{N} c_k f(n + k) = 0,$$

where we may suppose $c_N = 1$. It is known from the theory of difference equations, that in this case f is a linear combination of solutions of the form

$$n \mapsto n^j \lambda^n$$

where λ is a *characteristic root* of the equation, that is, $\sum_{k=0}^{N} c_k \lambda^k = 0$, and j is any nonnegative integer, smaller than the multiplicity of the root λ. This means, that the annihilator of μ contains exponentials, but the linear combinations of these exponentials are not necessarily dense. This is the case only if the characteristic polynomial of the difference equation has only simple roots. We observe, that the characteristic polynomial is equal to

$$\hat{\mu}(\lambda) = \int \lambda^{-n} d\mu(n),$$

which is an extension of the Fourier-transform of μ to the set of all exponentials. Then the closed ideal generated by μ is not the intersection of maximal ideals, but of ideals of the form

$$\{\mu : \hat{\mu}(\lambda) = \hat{\mu}'(\lambda) = \cdots = \hat{\mu}^{(j)}(\lambda) = 0\}$$

13

for some complex λ and nonnegative integer j. These ideals have the property, that they are contained in exactly one maximal ideal, hence they are primary ideals.

Hence it seems to be plausible to modify the spectral problems for $F(\mathbf{Z})$ in the following way:

i) Does every proper closed invariant subspace of $F(\mathbf{Z})$ contain an exponential?

ii) Are the linear combinations of the exponential monomials belonging to a proper closed invariant subspace of $F(\mathbf{Z})$ dense in this subspace?

For different choices of G and $F(G)$ these two questions have been dealt by several authors and have been answered either in the positive or in the negative. We shall deal mostly with the case $F(G) = \mathcal{C}(G)$, the space of all continuous complex valued functions.

Let G be any locally compact abelian group. A continuous homomorphism of G into the multiplicative group of nonzero complex numbers is called an *exponential*, or a *generalized character*. All exponentials on G form a group with respect to pointwise multiplication, and equipped with the topology of uniform convergence on compact sets this is a locally compact abelian group, which we call the *generalized character group* of G. Notation: $\tilde{G}$. In general $\tilde{G} \neq G$, but if G is compact, then $\tilde{G} = \hat{G}$.

The homomorphisms of G into the additive group of $\mathbf{C}$ or of $\mathbf{R}$ are called *additive functions*. *Monomials* are defined as products of additive functions, and by an *exponential monomial* we mean a function, which is the product of a monomial and an exponential. Hence monomials are the elements of the semigroup of functions generated by additive functions, and exponential monomials are elements of the semigroup generated by additive functions and exponentials. Linear combinations of monomials, resp. exponential monomials are called *polynomials*, resp. *exponential polynomials*. Hence polynomials are elements of the function algebra generated by additive functions, and exponential polynomials are elements of the function algebra generated by additive functions and exponentials. The general form of an exponential monomial is

$$x \mapsto a_1(x)^{\alpha_1} a_2(x)^{\alpha_2} \ldots a_n(x)^{\alpha_n} m(x),$$

where $a_1, a_2, \ldots a_n$ are additive, m is an exponential and $\alpha_1, \alpha_2, \ldots \alpha_n$ are nonnegative integers. Any exponential polynomial has the form

$$x \mapsto \sum_{i=1}^{N} p_i(a_1(x), a_2(x), \ldots a_n(x)) m_i(x),$$

14

where $a_1, a_2, \ldots a_n$ are additive functions, $m_1, m_2, \ldots m_N$ are exponentials and p_i is a complex polynomial in n variables ($i = 1, 2, \ldots N$). In general we suppose, that a maximal linearly independent set of real additive functions on G is fixed, and all exponential polynomials are built up from these additive functions.

The set of all continuous complex valued functions on G with the pointwise operations and with the topology of uniform convergence on compact sets is a locally convex topological vector space, which we denote by $\mathcal{C}(G)$. The dual of $\mathcal{C}(G)$ can be identified with the space of all *compactly supported complex Radon measures* on G, which is equipped with the weak $\star$ topology and is denoted by $\mathcal{M}_c(G)$. The pairing between $\mathcal{C}(G)$ and $\mathcal{M}_c(G)$ is given by

$$\langle f, \mu \rangle = \int f(x) d\mu(x)$$

for all f in $\mathcal{C}(G)$ and μ in $\mathcal{M}_c(G)$. Convolution between $\mathcal{C}(G)$ and $\mathcal{M}_c(G)$, further between $\mathcal{M}_c(G)$ and $\mathcal{M}_c(G)$ is defined in the usual way.

The Fourier-transform of a compactly supported continuous function f, resp., a compactly supported Radon measure μ has a natural extension to $\tilde{G}$, which is called the *Fourier-Laplace-transform*, and is given by

$$\hat{f}(m) = \int m(-x) f(x) dx,$$

resp.

$$\hat{\mu}(m) = \int m(-x) d\mu(x),$$

for any m in $\tilde{G}$.

A proper closed translation invariant subspace of $\mathcal{C}(G)$ will be called a *variety*. For any subset H of $\mathcal{C}(G)$ the *variety generated by H* is the smallest variety containing H and is denoted by $\tau(H)$. If, in particular, $H = \{f\}$, then we write $\tau(f)$ for $\tau(H)$. If $\tau(H)$, resp., $\tau(f)$ is not the whole $\mathcal{C}(G)$, then we call H, resp., f *mean periodic*. This means for f, that the linear combinations of the translates of f are not dense in $\mathcal{C}(G)$.

The annihilator sets will be used and will be denoted similarly, as before. Further, we have the respective theorem.

THEOREM. *The annihilator of a closed subspace in $\mathcal{C}(G)$ is a closed ideal if and only if the subspace is a variety.*

Hence we have a one-to-one correspondence between varieties of $\mathcal{C}(G)$ and closed proper ideals of $\mathcal{M}_c(G)$ again, and the basic spectral problems can be formulated equivalently either in $\mathcal{C}(G)$ or in $\mathcal{M}_c(G)$.

15

i) Does every variety in $\mathcal{C}(G)$ contain an exponential? - or equivalently, - is every proper closed ideal of $\mathcal{M}_c(G)$ contained in a maximal ideal?
ii) Are the linear combinations of exponential monomials in a variety in $\mathcal{C}(G)$ dense in this variety? - or equivalently, - is every proper closed ideal in $\mathcal{M}_c(G)$ the intersection of primary ideals?

By the theory of finite difference equations in the former example of $G = \mathbf{Z}$ the answer is affirmative for both questions. The first classical result in this respect is due to L.Schwartz ([SCZ1]).

THEOREM OF SCHWARTZ. *In $\mathcal{C}(\mathbf{R})$ any variety is the closed linear hull of the exponential monomials which are contained in it.*

In particular, any variety in $\mathcal{C}(\mathbf{R})$ contains an exponential. Using this fact, we can give all continuous - not necessarily bounded - solutions of the functional equation of d'Alembert, simply by repeating the argument used in the bounded case. L.Schwartz proved also the analogous result for the varieties of $E(\mathbf{R})$, the space of infinitely differentiable functions on $\mathbf{R}$ with the usual topology ([SCZ1]).

An extension of the result for $\mathbf{Z}^n$ is due to M.Lefranc ([LEF]).

THEOREM OF LEFRANC. *In $F(\mathbf{Z}^n)$ any variety is the closed linear hull of the exponential monomials which are contained in it.*

It turns out that an extension of the theorem of Schwartz for $\mathbf{R}^n$ is not possible if $n > 1$ ([SCZ1]). But a result of R.J.Elliot shows that this extension is possible for discrete abelian groups ([ELL2]).

THEOREM OF ELLIOT. *If G is a discrete abelian group, then in $\mathcal{C}(G)$ any variety is the closed linear hull of the exponential monomials which are contained in it.*

Although spectral synthesis does not hold for any variety in any locally compact abelian group, there are some special varieties which have spectral synthesis. These special varieties are characterized by the property, that their annihilator ideal is a *principal ideal*, which means that it is generated by a single measure. A classical result in this direction is due to B.Malgrange ([MGR]).

THEOREM OF MALGRANGE. *For any nonzero linear partial differential oper-*
ator $P(D)$ in $\mathbf{R}^n$ the linear hull of the exponential monomial solutions of the
partial differential equation $P(D)f = 0$ is dense in the set of all solutions.

For $n = 1$ this reduces to the well-known fact about homogeneous linear differential equations with constant coefficients, that the solutions are linear combinations of exponential monomial solutions. Here the basic space is
$E(\mathbf{R}^n)$, the space of infinitely differentiable functions on $\mathbf{R}^n$ with the usual
topology and $E(\mathbf{R}^n)^*$ is the space of all distributions with compact support.
The annihilator ideal is generated by the distribution $P(D)\delta$, where δ is the
Dirac-distribution.

The principal ideal technique was extended by L.Ehrenpreis ([EHR1],
[EHR2]) to obtain the following theorem.

THEOREM OF EHRENPREIS. *If the annihilator of a variety in $E(\mathbf{C}^n)$ is a princi-*
pal ideal, then the variety is the closed linear hull of the exponential monomials
which are contained in it.

The respective extension of this theorem for $\mathcal{C}(G)$, where G is a locally
compact abelian group is due to R.J.Elliot and J.E.Gilbert ([ELL1], [GIL1],
[GIL2]).

THEOREM OF ELLIOT-GILBERT. *If G is a locally compact abelian group, then*
in $\mathcal{C}(G)$ any variety, whose annihilator is a principal ideal, is the closed linear
hull of the exponential monomials which are contained in it.

Summarizing the results, we have spectral synthesis for any variety in
$\mathcal{C}(\mathbf{R})$ and for any variety in $\mathcal{C}(G)$ if G is a discrete abelian group, and we have
restricted spectral synthesis, that is spectral synthesis for those varieties in
$\mathcal{C}(G)$, whose annihilator ideal is principal, if G is any locally compact abelian
group.

In what follows we consider functional equations of the form

$$f * \mu = 0$$

where $f : G \to \mathbf{C}$ is an unknown function, and μ varies in a given subset Λ
of $\mathcal{M}_c(G)$. This is actually a system of convolution type functional equations.
The solution space of the above equation is obviously a closed linear translation
invariant subspace in $\mathcal{C}(G)$, that is, it is a variety, if $\Lambda \neq \{0\}$. Conversely, if a

variety V of $\mathcal{C}(G)$ is given, then by $V = (V^{\perp})^{\perp}$ we have, that V is the solution space of the system of functional equations

$$f * \mu = 0$$

for all $\mu \in V^{\perp}$. Hence the description of the varieties of $\mathcal{C}(G)$ is equivalent to the description of the solution spaces of systems of convolution type functional equations. By the spectral synthesis results on discrete groups the solution spaces of such equations are uniquely determined by the exponential monomial solutions. Hence a possible way to solve systems of convolution type equations is the following: first we determine all exponential monomial solutions, and then we try to identify the closed linear hull of the set of exponential monomial solutions. For the determination of the exponential monomial solutions the following lemma is useful ([SZÉ21]).

LEMMA. *Let* $V \subset \mathcal{C}(G)$ *be any variety,* $a_1, a_2, \ldots, a_n$ *linearly independent additive functions,* m *an exponential on* G *and let the nonnegative integers* $\alpha_1, \alpha_2, \ldots, \alpha_n$ *be given. If the exponential monomial*

$$x \mapsto a_1(x)^{\alpha_1} a_2(x)^{\alpha_2} \ldots a_n(x)^{\alpha_n} m(x)$$

belongs to V, *then the exponential monomials*

$$x \mapsto a_1(x)^{\beta_1} a_2(x)^{\beta_2} \ldots a_n(x)^{\beta_n} m(x)$$

belong to V *for any integers* $\beta_1, \beta_2, \ldots \beta_n$ *with* $0 \le \beta_i \le \alpha_i$ $(i = 1, 2, \ldots, n)$.

Hence in order to determine the exponential monomial solutions (that is, the spectral set of the equation), first it is useful to determine the exponential solutions. Obviously, an exponential m satisfies

$$m * \mu = 0$$

if and only if $\hat{\mu}(m) = 0$, hence the exponential solutions are just the common zeros of the Fourier-Laplace-transforms of the elements of Λ. We denote this set by $Z(\Lambda)$. Obviously, $\emptyset \neq Z(\Lambda) \subset \tilde{G}$. The next step should be to determine for all $m \in Z(\Lambda)$ those linearly independent additive functions $a_1, a_2, \ldots, a_n$, for which $a_1 a_2 \ldots a_n m$ is a solution, but here also necessary to have, that $a_i m$ is a solution for $i = 1, 2, \ldots, n$. Continuing this process it may turn out - depending on the special form of the system of equations - that a more explicit characterization of the spectral set can be given, which may yield a complete characterization of the solutions. We shall see several examples in the subsequent sections which illustrate the method in particular cases.

18

CHAPTER 1

POLYNOMIALS AND EXPONENTIAL POLYNOMIALS

1. Multi-additive functions on semigroups

The study of polynomials on semigroups is based on the notion of multi-additive functions. Here we collect some basic algebraic facts about multi-additive functions on commutative semigroups.

Let G, S be commutative semigroups, n a positive integer and let $A : G^n \to S$ a function. We call A *n-additive* if it is a homomorphism of G into S in each variable. If $n = 1$ or 2, then we call A *additive* or *biadditive*, respectively. We extend this terminology also for the case $n = 0$ by letting $G^0 = G$ and calling 0-additive any constant function from G to S. We call A *multi-additive* if there exists a nonnegative integer n such that A is n-additive. The *diagonalization* of the n-additive function $A : G^n \to S$ is denoted by A^*, that is,

$$A^*(x) = A([x]_n)$$

for any x in G. Here $[x]_n$ denotes the element of G^n all components of which are equal to x. Evidently $A^* = A$ for $n = 0$ or 1.

It is obvious that for a fixed nonnegative integer n, all n-additive functions from G into S form a commutative semigroup. Moreover, we have the following theorem, which is very easy to prove by definition.

THEOREM 1.1. *Let G, S be commutative semigroups and n a nonnegative integer. All n-additive functions from G into S form a*
 (i) commutative group, if S is a commutative group;
 (ii) module over the ring R, if S is a module over R;
(iii) linear space over the field F, if S is a linear space over F.

From the definition of multi-additive functions it follows that, if n is any positive integer, then each n-additive function $A : G^n \to S$ satisfies

$$A(x_1, x_2, \ldots, x_{i-1}, kx_i, x_{i+1}, \ldots, x_n) = kA(x_1, x_2, \ldots, x_{i-1}, x_i, x_{i+1}, \ldots, x_n)$$

for $i = 1, 2, \ldots, n$, whenever $x_1, x_2, \ldots, x_n$ are in G and k is a positive integer. Further, the same property holds for any integer k, if G and S are groups, and for any rational k, if G and S are linear spaces over the rationals (or over any field of characteristic zero). Respectively, for the diagonalization of A we have

$$A^*(kx) = k^n A^*(x)$$

for any x in G.

For symmetric multi-additive functions the following *binomial theorem* holds.

LEMMA 1.2. *Let G, S be commutative semigroups and n a nonnegative integer. If $A : G^n \to S$ is n-additive and symmetric, then we have for all x, y in G*

$$A^*(x + y) = \sum_{k=0}^{n} A([x]_k, [y]_{n-k}).$$

PROOF: Let $n \geq 1$, then we have for all x, y in G

$$A^*(x + y) = A(x + y, [x + y]_{n-1}) = A(x, [x + y]_{n-1}) + A(y, [x + y]_{n-1}).$$

As the function $(x_1, x_2, \ldots, x_{n-1}) \mapsto A(z, x_1, x_2, \ldots, x_{n-1})$ is evidently $n - 1$-additive and symmetric for any fixed z in G, and its diagonalization is the function $x \mapsto A(z, [x]_{n-1})$, the above equation gives the statement by induction as follows

$$A^*(x + y) = \sum_{k=0}^{n-1} \binom{n-1}{k} \left[A(x, [x]_k, [y]_{n-1-k}) + A(y, [x]_k, [y]_{n-1-k}) \right] =$$

$$= \sum_{k=0}^{n-1} \binom{n-1}{k} \left[A([x]_{k+1}, [y]_{n-k-1}) + A([x]_k, [y]_{n-k}) \right] =$$

$$= \sum_{k=1}^{n} \binom{n-1}{k-1} A([x]_k, [y]_{n-k}) + \sum_{k=0}^{n-1} \binom{n-1}{k} A([x]_k, [y]_{n-k}) =$$

$$= A^*(x) + \sum_{k=1}^{n-1} \left[\binom{n-1}{k-1} + \binom{n-1}{k} \right] A([x]_k, [y]_{n-k}) + A^*(y) =$$

$$= \sum_{k=0}^{n} \binom{n}{k} A([x]_k, [y]_{n-k}).$$

This implies also the *polynomial theorem* as follows:

LEMMA 1.3. *Let G, S be commutative semigroups and n a nonnegative integer. If $A : G^n \to S$ is n-additive and symmetric then we have for any positive integer m and for all $x_1, x_2, \ldots, x_m$ in G*

$$A^*(x_1 + x_2 + \cdots + x_m) =$$

$$= \sum_{k_1+k_2+\cdots+k_m=n} \frac{n!}{k_1! k_2! \ldots k_m!} A([x]_{k_1}, [x]_{k_2}, \ldots, [x]_{k_m}).$$

The *polarization formula* for multi-additive symmetric functions is expressed by the following lemma.

LEMMA 1.4. *Let G be a commutative semigroup, S a commutative group and n a nonnegative integer. If $A : G^n \to S$ is n-additive and symmetric, then we have for all $y_1, y_2, \ldots, y_m$ in G*

$$\Delta_{y_1, y_2, \ldots, y_m} A^* = \begin{cases} 0 & \text{for } m > n, \\ n! A(y_1, y_2, \ldots, y_n) & \text{for } m = n. \end{cases}$$

PROOF: Obviously the statement for $m = n$ implies that for $m > n$. We prove the lemma by induction on n, and it is trivial for $n = 1$. Supposing that it has been proved for all values not greater than n, we prove it for $n + 1$. By the properties of the difference operators and by Lemma 1.2 we have for all x in G

$$\Delta_{y_1, y_2, \ldots, y_n, y_{n+1}} A^*(x) = \Delta_{y_1, y_2, \ldots, y_n} [\Delta_{y_{n+1}} A^*](x) =$$

$$= \Delta_{y_1, y_2, \ldots, y_n} \left[\sum_{k=0}^{n} \binom{n+1}{k} A([x]_k, [y_{n+1}]_{n+1-k}) \right] =$$

$$= \sum_{k=0}^{n} \binom{n+1}{k} \Delta_{y_1, y_2, \ldots, y_n} A([x]_k, [y_{n+1}]_{n+1-k}) =$$

$$= \sum_{k=1}^{n} \binom{n+1}{k} \Delta_{y_1, y_2, \ldots, y_n} A([x]_k, [y_{n+1}]_{n+1-k}) =$$

$$= (n+1)n! A(y_1, y_2, \ldots, y_n, y_{n+1}) = (n+1)! A(y_1, y_2, \ldots, y_n, y_{n+1}),$$

as

$$x \mapsto A([x]_k, [y_{n+1}]_{n+1-k})$$

is the diagonalization of the k-additive and symmetric function

$$(x_1, x_2, \ldots, x_k) \mapsto A(x_1, x_2, \ldots, x_k, y_{n+1}, \ldots, y_{n+1})$$

for $k = 1, 2, \ldots, n$.

COROLLARY 1.5. *Let G be a commutative semigroup, S a commutative group and n a nonnegative integer. If $A : G^n \to S$ is n-additive and symmetric, then we have for all x, y in G*

$$\Delta_y^n A^*(x) = n! A^*(y).$$

The distinguished role played by the symmetric ones among all multi-additive functions is also explained by the following lemma.

LEMMA 1.6. *Let G be a commutative semigroup, S a commutative group and n a nonnegative integer. Let the multiplication by $n!$ be either surjective in G or injective in S. Then, for any $A : G^n \to S$ n-additive and symmetric function, $A^* = 0$ implies $A = 0$.*

PROOF: If $x \mapsto n!x$ is injective in S, then the statement is a direct consequence of Lemma 1.4. In the other case we also observe, that

$$n! A(y_1, y_2, \ldots, y_n) = A(n! y_1, y_2, \ldots, y_n).$$

We note, that in the last three lemmata the condition on the symmetry of A cannot be omitted. Let for instance $G = \mathbf{R}$ and let H be a basis of $\mathbf{R}$, considered as a $\mathbf{Q}$-linear space (a *Hamel-basis* of $\mathbf{R}$). We denote the canonical projection of $\mathbf{R}$ onto $\mathbf{Q}$ corresponding to the basis element h by $\mathbf{p}_h$, that is, we have for any x in $\mathbf{R}$ and for any h in H

$$x = \sum_{h \in H} \mathbf{p}_h(x) x.$$

Let A be a nonzero real function on H^2 with $A(h_1, h_2) = -A(h_2, h_1)$ for any h_1, h_2 in H. We extend A to a biadditive function on $\mathbf{R}^2$ by setting

$$A(x_1, x_2) = \sum_{h_1, h_2 \in H} \mathbf{p}_{h_1}(x_1) \mathbf{p}_{h_2}(x_2) A(h_1, h_2)$$

for any x_1, x_2 in $\mathbf{R}$. Obviously A is antisymmetric, hence $A^* = 0$, but A is nonzero.

REFERENCES 1.7.

The fundamental properties of multi-additive functions on abelian groups (similarly to those of the multi-linear functions on linear spaces) had been extensively studied in [FRE4], [LIJ3] and [MAZ]. For various results and for further references concerning multi-additive functions see also [ACZ7], [DJO3], [GAJ2], [GAJ4], [GHE3], [KCZ4], [SZÉ1], [SZÉ3], [SZÉ7], [SZÉ8], [SZÉ20].

2. Polynomials on semigroups

The theory of polynomial operations on semigroups, groups and linear spaces has its origin in [FRE4], [LIJ3], [MAZ]. The definition is due to Fréchet and Banach (see [FRE4], [MAZ]), while [MAZ] is devoted to a detailed study of polynomial operations on linear spaces. The basic theorems on the canonical representation of polynomials on abelian groups can be found also in [DJO3], [LIJ3].

Let G, S be commutative topological semigroups. A continuous function $p : G \to S$ will be called a *polynomial* from G into S, if it has a representation as the sum of diagonalizations of multi-additive functions from G into S. Hence polynomials are the continuous elements of the semigroup generated by the diagonalizations of multi-additive functions from G into S. In other words, a continuous function $p : G \to S$ is a polynomial if and only if it has a representation

$$p = \sum_{k=0}^{n} A_k^*,$$

where n is a nonnegative integer and $A_k : G \to S$ is a k-additive function ($k = 0, 1, \ldots, n$). In this case we also say that p is a *polynomial of degree at most n*. If S is a group and p is a polynomial of degree at most n, then obviously $\Delta_y p$ is a polynomial of degree at most $n - 1$ for any y in G. Polynomials of the form A^*, where A is any multi-additive function, are called *homogeneous polynomials*. The set of all polynomials from G into S will be denoted by $\mathcal{P}(G, S)$, and $\mathcal{P}_n(G, S)$ denotes the set of polynomials of degree at most n. Evidently, $\mathcal{P}(G, S)$ and $\mathcal{P}_n(G, S)$ are translation invariant. If we equip $\mathcal{P}(G, S)$ with the compact-open topology of $\mathcal{C}(G, S)$, then obviously $\mathcal{P}(G, S)$ is a commutative topological semigroup.

We note that by our definition, the notion of a polynomial from G into S (and hence also the meaning of $\mathcal{P}(G, S)$) depends heavily on the topologies of G and S. Obviously, by changing any of these topologies, $\mathcal{P}(G, S)$ may change as well. The largest class of polynomials from G into S is the one, which occurs when G is given the discrete topology. The elements of this class will be called *algebraic polynomials* from G to S. In what follows, if no topology on G is mentioned, then we always refer to the discrete topology of G, and by "polynomials" on G we mean the algebraic polynomials. Nevertheless, in some cases, if different topologies on the same underlying semigroup will be considered, there should be made a strict distinction - also in the notation - between the respective - maybe different - notions of polynomials. Thus, for instance, when considering $\mathbf{R}$ with the discrete topology, we write $\mathbf{R}_d$ instead of $\mathbf{R}$, and the respective set of polynomials will be denoted by $\mathcal{P}(\mathbf{R}_d, S)$. Similarly, if G is any commutative topological semigroup, then $\mathcal{P}(G_d, S)$ denotes the set of all algebraic polynomials from G into S. In this case the topology of $\mathcal{P}(G_d, S)$ is that of pointwise convergence.

The following theorem is an immediate consequence of Theorem 1.1.

THEOREM 2.1. *Let G, S be commutative topological semigroups and n a non-negative integer. The set of all polynomials of degree at most n, resp., the set of all polynomials from G into S forms a*
(i) *topological abelian group, if S is a topological group;*
(ii) *topological module over the topological ring R, if S is a topological module over R;*
(iii) *topological linear space over the topological field K, if S is a topological linear space over K.*

THEOREM 2.2. *Let G be a commutative topological semigroup and R a topological ring. The set of all polynomials from G into the additive topological group of R forms a*
(i) *(commutative) topological ring, if R is a (commutative) topological ring;*
(ii) *(commutative) topological algebra over the topological field K, if R is a (commutative) topological algebra over K.*

PROOF: Using the previous theorems, it is enough to prove, that the pointwise product of two polynomials is a polynomial. But this follows from the elementary fact, that the product of the diagonalizations of two multi-additive functions is just the diagonalization of the tensorial product of the two multi-additive functions.

THEOREM 2.3. *Let G be a commutative semigroup, S a commutative group, n a nonnegative integer and suppose that multiplication by $n!$ is bijective either on G or on S. Then any polynomial $p : G \rightarrow S$ of degree at most n has a*

25

unique representation in the form

$$p = \sum_{k=0}^{n} A_k^*,$$

where $A_k : G^k \to S$ is a k-additive and symmetric function ($k = 0, 1, \ldots, n$), further $A_n \neq 0$, whenever p is not of degree at most $n - 1$.

PROOF: In both cases it is enough to show, that for any $A : G^k \to S$ k-additive function there exists an $\tilde{A} : G^k \to S$ k-additive and symmetric function with $\tilde{A}^* = A^*$ ($k = 0, 1, \ldots, n$); then our statement follows from Lemma 1.3. We define

$$\tilde{A}(x_1, x_2, \ldots, x_k) = \begin{cases} \sum A(\frac{1}{k!} x_{\sigma(1)}, x_{\sigma(2)}, \ldots, x_{\sigma(k)}) \\ \frac{1}{k!} \sum A(x_{\sigma(1)}, x_{\sigma(2)}, \ldots, x_{\sigma(k)}) \end{cases}$$

(the sums are taken for all permutations σ of the set $\{1, 2, \ldots, k\}$) respectively to the cases, where multiplication by $n!$ is bijective either on G or on S. Then A satisfies all requirements and the theorem is proved.

Under the assumptions of Theorem 2.3 the given representation is called the *canonical representation* of p, and we call A_k^* its *homogeneous term of degree k*; further A_n^* is called its *leading term*, whenever it is not identically zero. In this case we define the degree of the polynomial p as $\deg p = n$. The degree of nonzero constant polynomials is defined as 0, and the degree of the zero polynomial is defined as -1. It follows, that two polynomials are equal if and only if their homogeneous terms of the same degree are all equal.

We note that the conditions of Theorem 2.3 are obviously satisfied, if S is a linear space over a field. In particular, the elements of $\mathcal{P}(G, \mathbf{R})$, resp. $\mathcal{P}(G, \mathbf{C})$, are called real, resp. complex polynomials. We write simply $\mathcal{P}(G)$ for $\mathcal{P}(G, \mathbf{C})$ and $\mathcal{P}_n(G)$ for $\mathcal{P}_n(G, \mathbf{C})$.

THEOREM 2.4. *Let G be a commutative topological semigroup, X a complex topological vector space and n a nonnegative integer. Then $\mathcal{P}_n(G, X)$ is a closed linear subspace of $\mathcal{P}(G, X)$.*

PROOF: It is enough to show, that, if p is the limit in $\mathcal{P}(G, X)$ of a convergent net $\{p_\alpha\}$ of polynomials of degree at most n, then p is itself a polynomial of degree at most n. Let p_α have the canonical representation

$$p_\alpha = \sum_{k=0}^{n} A_{\alpha, k}^*,$$

26

then we obtain by Lemma 1.4 for all $x, y_1, y_2, \ldots, y_n$ in G

$$A_{\alpha,n}(y_1, y_2, \ldots, y_n) = \frac{1}{n!} \Delta_{y_1, y_2, \ldots, y_n} p_\alpha(x) =$$

$$= \frac{1}{n!} \sum_{0 \le i_1 < \cdots < i_k \le n} (-1)^{n-k} p_\alpha(x + y_{i_1} + y_{i_2} + \cdots + y_{i_k}).$$

This shows, that the net $\{A_{\alpha,n}\}$ is uniformly convergent on any compact subset of G^n, and its limit A_n is obviously an n-additive and symmetric function. Repeating this argument for $p_\alpha - A^*_{\alpha,n}$ instead of p_α, etc., we have, that all the nets $\{A_{\alpha,k}\}$ are uniformly convergent on all compact sets in G^k, and the limit $A_k = \lim_\alpha A_{\alpha,k}$ is a k-additive and symmetric function for all $k = 0, 1, \ldots, n$. Then it follows

$$\sum_{k=0}^{n} A^*_k = \lim_\alpha \sum_{k=0}^{n} A^*_{\alpha,k} = \lim_\alpha p_\alpha = p,$$

which proves our statement.

We note, that we have actually proved the following statement: if the net of polynomials of degree at most n converges uniformly on all compact sets, then the nets of their homogeneous terms of degree k converge uniformly on all compact sets, for $k = 0, 1, \ldots, n$.

The following theorem is useful concerning functional equations.

THEOREM 2.5. *Let G, S be commutative semigroups, n a nonnegative integer, $\varphi_i, \psi_i : G \to G$ homomorphisms ($i = 1, 2, \ldots, n+2$) and suppose that multiplication by $n!$ is bijective on S. Then the polynomials $p_i : G \to S$ ($i = 1, 2, \ldots, n+2$) of the form $p_i = \sum_{i=1}^{n} A^*_{i,k}$, where $A_{i,k} : G \to S$ is a k-additive and symmetric function ($k = 0, 1, \ldots, n; i = 1, 2, \ldots, n+2$), satisfy the functional equation*

$$\sum_{i=1}^{n+2} p_i(\varphi_i(x) + \psi_i(y)) = 0$$

for all x, y in G if and only if the following relations hold

$$\sum_{i=1}^{n+2} A_{i,k}([\varphi_i(x)]_j, [\psi_i(y)]_{k-j}) = 0$$

for every x, y in G, and for $j = 0, 1, \ldots, n; k = j, j+1, \ldots, n$.

PROOF: If the given polynomials satisfy our equation, then by substitution and by Lemma 1.2 we have

$$\sum_{i=1}^{n+2} \sum_{k=0}^{n} \sum_{j=0}^{k} \binom{k}{j} A_{i,k}([\varphi_i(x)]_j, [\psi_i(y)]_{k-j}) = 0.$$

Here the function $x \mapsto A_{i,k}([\varphi_i(x)]_j, [\psi_i(y)]_{k-j})$ is a homogeneous polynomial of degree j for every fixed y in G, hence we have by Theorem 2.3

$$\sum_{i=1}^{n+2} \sum_{k=j}^{n} \binom{k}{j} A_{i,k}([\varphi_i(x)]_j, [\psi_i(y)]_{k-j}) = 0$$

for every x, y in G, and for $j = 0, 1, \ldots, n$. Repeating this argument we obtain

$$\sum_{i=1}^{n+2} A_{i,k}([\varphi_i(x)]_j, [\psi_i(y)]_{k-j}) = 0$$

for all x, y in G and for $j = 0, 1, \ldots, n; \ k = j, j+1, \ldots, n$, as multiplication by $n!$ is bijective in S. To prove the converse, we multiply our last equation by $\binom{k}{j}$, and add the equations for $k = j, j+1, \ldots, n$, and for $j = 0, 1, \ldots, n$.

Now we introduce a translation invariant subclass of $\mathcal{P}(G)$. If m is a positive integer, $P : \mathbf{R}^m \to \mathbf{C}$ is a (classical) polynomial in m variables and $a_k : G \to \mathbf{R}$ $(k = 1, 2, \ldots, m)$ are continuous additive functions, then the function

$$x \mapsto P(a_1(x), a_2(x), \ldots, a_m(x))$$

is obviously a polynomial. We call complex polynomials of this form *normal polynomials*. In other words, normal polynomials on a commutative topological semigroup are linear combinations of products of continuous real homomorphisms. If the polynomial P above is a monomial, then we call the respective normal polynomial a *monomial* too. That is, monomials on a commutative topological semigroup are products of continuous real homomorphisms, and any normal polynomial is a complex linear combination of monomials.

The following theorem is very easy to prove by definition.

THEOREM 2.6. *The set of all normal polynomials on a commutative topological semigroup is the function algebra generated by all continuous complex homomorphisms of the semigroup.*

Dealing with normal polynomials, it is very useful to apply the multi-index notation. We recall, that elements of $\mathbf{N}^n$ for any positive integer n are called (n-dimensional) *multi-indices*. Addition, multiplication and inequalities between multi-indices of the same dimension are defined componentwise. Further, we define x^α for any n-dimensional multi-index α, and for any $x = (x_1, x_2, \ldots, x_n)$ in $\mathbf{R}^n$ by

$$x^\alpha = x_1^{\alpha_1} x_2^{\alpha_2} \ldots x_n^{\alpha_n}$$

with $0^0 = 0$. We also use the notation $|\alpha| = \alpha_1 + \alpha_2 + \ldots \alpha_n$. With these notation any normal polynomial of degree at most N on the commutative topological semigroup G has the form

$$p(x) = \sum_{|\alpha| \leq N} c_\alpha a(x)^\alpha \ .$$

for x in G, where $a = (a_1, a_2, \ldots, a_n)$ is a continuous additive function from G into $\mathbf{R}^n$. The homogeneous term of degree k of p is $\sum_{|\alpha|=k} c_\alpha a(x)^\alpha$.

LEMMA 2.7. *Let G be a commutative topological semigroup, n a positive integer and let $a = (a_1, a_2, \ldots, a_n)$, where $a_1, a_2, \ldots, a_n$ are linearly independent continuous real additive functions on G. Then the monomials $\{a^\alpha\}$ for different multi-indices α are linearly independent.*

PROOF: By Theorem 2.3 it is enough to show that an equation of the form

$$\sum_{|\alpha|=N} c_\alpha a(x)^\alpha = 0$$

for all x in G implies that $c_\alpha = 0$ for all α. If $N = 0$, then this is trivial, hence we suppose $N \geq 1$ and we put $x + y$ for x in the above equation. Then, by the Taylor-formula we have

$$\sum_{|\alpha|=N} \sum_{\beta \leq \alpha} c_\alpha \lambda_\beta a(x)^{\alpha-\beta} a(y)^\beta = 0$$

for all x, y in G, where the λ's are nonzero complex numbers. Here the right hand side is a polynomial in x, which is identically zero, thus, by Theorem 2.3

29

all homogeneous terms of the same degree of it must be zero. For the terms of the first degree this means

$$\sum_{|\alpha|=N} c_\alpha \sum_{i=1}^{n} \lambda_{\alpha[i]} a_i(x) a(y)^{\alpha[i]} = 0$$

for all x, y in G, where $\alpha[i] = (\alpha_1, \alpha_2, \ldots, \alpha_{i-1}, \alpha_i - 1, \alpha_{i+1}, \ldots, \alpha_{n-1}, \alpha_n)$ (if here $\alpha_i = 1$, then $\lambda_{\alpha[i]} = 0$). By the linear independence of $a_1, a_2, \ldots, a_n$ it follows

$$\sum_{|\alpha|=N} c_\alpha \lambda_{\alpha[i]} a(y)^{\alpha[i]} = 0$$

for all y in G. As $|\alpha[i]| = N - 1$ $(i = 1, 2, \ldots, n)$, our statement follows by induction.

LEMMA 2.8. *Let G be a commutative topological semigroup, n a positive integer, α an n-dimensional multi-index and let $a = (a_1, a_2, \ldots, a_n)$, where $a_1, a_2, \ldots, a_n$ are linearly independent continuous real additive functions on G. Then the smallest translation invariant linear space of continuous complex valued functions on G, which contains the monomial a^α has a basis consisting of all the monomials a^α with $\beta \leq \alpha$. (Consequently, its dimension is equal to $(\alpha_1 + 1)(\alpha_2 + 1)\ldots(\alpha_n + 1)$.)*

PROOF: By the Taylor-formula we have for all x, y in G

$$a(x + y)^\alpha = \sum_{\beta \leq \alpha} \lambda_\beta a(x)^{\alpha - \beta} a(y)^\beta,$$

with some nonzero complex constants λ_β. Then, on the one hand, any translate of a^α is a linear combination of monomials of the form a^β with $\beta \leq \alpha$, and on the other hand, by the previous lemma all these monomials are linear combinations of translates of a^α. Hence our statement is a consequence of Lemma 2.7.

The last two lemmata show that concerning monomials and normal polynomials on an abelian semigroup it is useful to assume, that they all are built up from a fixed linearly independent set of real additive functions. We shall do this in the sequel.

REFERENCES 2.9.

For further references and results concerning polynomials on semigroups see [ACZ6], [ACZ7], [ACZ8], [ACZ13], [ACZ14], [ACZ15], [ACZ16], [ACZ18],

[ALE], [ANG1], [ANG2], [ING], [FRE1], [FRE2], [FRE3], [GAJ2], [GAJ4], [GER1], [GHE1], [GHE3], [HAS4], [HOS1], [HOS2], [KCZ4], [LIJ1], [LIJ2], [MAR], [MCK1], [POP1], [POP2], [REI1], [REI2], [SZÉ1], [SZÉ3], [SZÉ7], [SZÉ8], [SZÉ12], [SZÉ22].

3. Regularity properties of polynomials

The definition of polynomials requires continuity. In this section we show that in general mild regularity conditions on algebraic polynomials imply their continuity. This property of special algebraic polynomials has been observed by several authors and has been proved in several cases. The first classical results in this direction concerned mostly with additive functions on topological groups. In the case of additive functions on the real line it has been proved that, for instance, continuity at one point, monotonicity on an arbitrary interval, boundedness or one-sided boundedness on an arbitrary interval, measurability on a measurable set of positive measure, etc., imply that an additive function must be continuous everywhere. These results have been extended respectively for multi-additive functions, and for algebraic polynomials. In this section we collect some global and local results of this type, which will be useful for our later purposes.

THEOREM 3.1. *Let G be a commutative semigroup and let X be a locally convex topological vector space. Then any bounded polynomial from G into X is constant.*

PROOF: Let $p = \sum_{k=0}^{n} A_k^*$ be the canonical representation of p. We show that $A_n^* = 0$ for $n > 0$. By Lemma 1.4 we have

$$A_n(y_1, y_2, \ldots, y_n) = \frac{1}{n!} \Delta_{y_1, y_2, \ldots, y_n} p(x) =$$

$$= \frac{1}{n!} \sum_{0 \le i_1 < i_2 < \cdots < i_k \le n} (-1)^{n-k} p(x + y_{i_1} + y_{i_2} + \cdots + y_{i_k})$$

for all $x, y_1, y_2, \ldots, y_n$ in G, which shows that A_n is bounded. On the other hand, the n-additivity of A_n implies for $n > 0$

$$A_n^*(my) = m^n A_n^*(y),$$

whenever y is in G and m is any positive integer. Suppose, that $A_n^*(x_0) \neq 0$ for some x_0 in G. We choose a balanced and absorbing neighborhood W of zero in X, then there is a real α for which

$$m^n A_n^*(x_0) = A_n^*(mx_0) \in \alpha W$$

for all positive integers m. Then $\alpha m^{-n} < 1$ for some m and we have

$$A_n^*(x_0) = m^{-n} A_n^*(mx_0) \in \alpha m^{-n} W \subseteq W,$$

which is a contradiction, hence the theorem is proved.

THEOREM 3.2. *Let G be a topological abelian group which is generated by any neighborhood of zero and let X be a linear space. If an algebraic polynomial from G into X vanishes on a nonvoid open set, then it vanishes everywhere.*

PROOF: Using the same notations as in the previous theorem we prove that $A_n = 0$. As any translate of an algebraic polynomial is an algebraic polynomial again, we may suppose that p vanishes on a neighborhood U of zero. We choose a neighborhood V of zero such that $x + y_1 + \cdots + y_n \in U$ whenever $x, y_1, \ldots, y_n \in V$. Then we obtain by Lemma 1.4

$$A_n(y_1, y_2, \ldots, y_n) = \frac{1}{n!} \Delta_{y_1, y_2, \ldots, y_n} p(x) =$$

$$= \frac{1}{n!} \sum_{0 \leq i_1 < i_2 < \cdots < i_k \leq n} (-1)^{n-k} p(x + y_{i_1} + y_{i_2} + \cdots + y_{i_k}) = 0,$$

whenever $x, y_1, \ldots, y_n \in V$, that is, A_n vanishes on V^n. Using the n-additivity of A_n and the fact that V^n generates G, our statement follows.

THEOREM 3.3. *Let G be a locally compact abelian group which is generated by any neighborhood of zero and let X be a linear space. If an algebraic polynomial from G into X vanishes on a measurable set of positive measure, then it vanishes everywhere.*

PROOF: Using the same notations as in the previous theorem we prove that $A_n^* = 0$. Let $K \subseteq G$ be a compact set of positive Haar-measure $\lambda(K) > 0$ such that p vanishes on K. It is well-known [KUR1] that the function $x \mapsto \lambda(K \cap K - x \cap \cdots \cap K - nx)$ is continuous on G, and as its value is positive at zero, we have that there is a neighborhood $U \subseteq G$ of zero for which $y \in U$

33

implies $\lambda(K \cap K - y \cap \cdots \cap K - ny) > 0$. That is, for all y in U there exists an x in K such that $x + ky \in K$ for $k = 1, 2, \ldots, n$. Then we have by Lemma 1.4 and by Corollary 1.5

$$A_n^*(y) = \frac{1}{n!}\Delta_y^n p(x) = \frac{1}{n!}\sum_{k=0}^{n}(-1)^{n-k}p(x + ky) = 0,$$

whenever $y \in U$. This means that the algebraic polynomial A_n^* vanishes on U, and hence, by the previous theorem, it vanishes everywhere.

THEOREM 3.4. Let G be a topological abelian group and let X be a topological vector space. Then all multi-additive functions in the canonical representations of polynomials from G into X are continuous.

PROOF: This is an easy consequence of Lemma 1.4.

THEOREM 3.5. Let G, S be commutative topological semigroups and $a : G \to S$ an additive function. If either G or S is a group and a is continuous at a point, then it is continuous everywhere.

PROOF: The statement follows from the identities

$$a(x + y) = a(x_0 + y) + a(x - x_0),$$

or

$$a(x + y) = a(x_0 + y) + a(x) - a(x_0)$$

which hold for all x, x_0, y in G, respectively to the cases if G or S is a group.

THEOREM 3.6. Let G be a topological abelian group which is generated by any neighborhood of zero and let X be a locally convex topological vector space. Then any algebraic polynomial from G into X, which is continuous at a point, is continuous everywhere.

PROOF: Using the notations of Theorem 3.1, the formula

$$A_n(y_1, y_2, \ldots, y_n) = \frac{1}{n!}\Delta_{y_1, y_2, \ldots, y_n} p(x) =$$

$$= \frac{1}{n!}\sum_{0 \leq i_1 < i_2 < \cdots < i_k \leq n}(-1)^{n-k}p(x + y_{i_1} + y_{i_2} + \cdots + y_{i_n}),$$

which holds for all $x, y_1, y_2, \ldots, y_n$ in G, shows that A_n is continuous at the point $(0, 0, \ldots, 0)$. It is enough to prove that A_n is continuous on G^n and we may suppose that $n \geq 1$. First we show that for all $k = 1, 2, \ldots, n$ and $x_{k+1}, x_{k+2}, \ldots, x_n \in G$ the function

$$(g_1, g_2, \ldots, g_k) \mapsto A_n(g_1, g_2, \ldots, g_k, x_{k+1}, x_{k+2}, \ldots, x_n)$$

is continuous at the zero (of G^k). Let $W \subseteq X$, $U \subseteq G$ be neighborhoods of zero with W is convex and $A_n(U, U, \ldots, U) \subseteq W$. As G is generated by U, there exists a positive integer N, such that $x_{k+1}, x_{k+2}, \ldots, x_n \in NU$. Further there exists a neighborhood $V \subseteq G$ such that $N^{n-k}V \subseteq U$. Let $g_1, g_2, \ldots, g_k \in V$, then $N^{n-k}g_1, g_2, \ldots, g_k \in U$ and

$$A_n(g_1, g_2, \ldots, g_k, x_{k+1}, x_{k+2}, \ldots, x_n) =$$

$$= \frac{1}{N^{n-k}} A_n(N^{n-k}g_1, g_2, \ldots, g_k, y_1^{(k+1)} + \cdots + y_N^{(k+1)}, \ldots, y_1^{(n)} + \cdots + y_N^{(n)}),$$

where $y_1^{(i)}, y_2^{(i)} \ldots, y_N^{(i)} \in U$ $(i = k+1, k+2, \ldots, n)$. By the n-additivity of A_n the right hand side is a convex combination of elements of W, hence, it belongs to W, which proves our first statement.

Now let $x_1, x_2, \ldots, x_n \in G$ be arbitrary. Then, for any $g_1, g_2, \ldots, g_n$ in G the difference

$$A_n(x_1 + g_1, x_2 + g_2, \ldots, x_n + g_n) - A_n(x_1, x_2, \ldots, x_n)$$

can be expressed as a sum of terms which belong to an arbitrary given neighborhood of zero in X, whenever $g_1, g_2, \ldots, g_n$ are chosen from an appropriate neighborhood of zero in G, by the statement proved above. This implies the continuity of A_n at $x_1, x_2, \ldots, x_n$.

THEOREM 3.7. *Let G be a topological abelian group which is generated by any neighborhood of zero and let X be a locally convex topological vector space. Then any algebraic polynomial from G into X, which is bounded on a nonvoid open set, is continuous.*

PROOF: Using the above notations, one sees that A_n is bounded on a neighborhood of zero in G. Let $U \subseteq G$ be a neighborhood of zero for which $A_n(U, U, \ldots, U)$ is bounded in X. It is enough to prove that A_n^* is continuous at zero. Supposing the contrary, there exists a convex balanced neighborhood $W \subseteq X$ of zero such that every neighborhood of zero in G contains an element

35

z with $A_n^*(z) \notin W$. On the other hand, there exists a positive integer N such that $A_n(U, U, \ldots, U) \subseteq NW$. Let $m > \sqrt[n]{N}$ be an integer and let $V \subseteq G$ be a neighborhood of zero for which $mV \subseteq U$ and $z \in V$ such that $A_n^*(z) \notin W$. Then $mz \in U$, and hence $A_n^*(mz) \in NW$, but $A_n^*(mz) = m^n A_n^*(z) \notin m^n W$. On the other hand, $m^n W \supseteq NW$, which is a contradiction and our theorem is proved.

THEOREM 3.8. *Let G be a locally compact abelian group which is generated by any neighborhood of zero and let X be a locally convex topological vector space. Then any algebraic polynomial from G into X, which is bounded on a measurable set of positive measure, is continuous.*

PROOF: We use the notations of Theorem 3.1. By the conditions p is bounded on a compact set $K \subseteq G$ with $\lambda(K) > 0$. Let $U \subseteq G$ be a neighborhood of zero, for which $x \in U$ implies $K \cap K - x \cap \cdots \cap K - nx \neq \emptyset$ (see the proof of Theorem 3.3). Then there exists an $y \in G$ such that $y, y + x, \ldots, y + nx \in K$. We have

$$A_n^*(x) = \frac{1}{n!} \Delta_x^n p(y) = \frac{1}{n!} \sum_{k=0}^{n} \binom{n}{k} (-1)^{n-k} p(y + kx),$$

and this means that A_n is bounded on U. Then by Theorem 3.7 our statement follows.

THEOREM 3.9. *Let G be a locally compact abelian group which is generated by any neighborhood of zero and let X be a locally convex and locally bounded topological vector space. Then any algebraic polynomial from G into X, which is measurable on a measurable set of positive measure, is continuous.*

PROOF: We use the above notations. By the conditions p is measurable on a compact set $K \subseteq G$ with $\lambda(K) > 0$. Let $W \subseteq X$ be a bounded, balanced and absorbing neighborhood of zero and let

$$L_n = \{x : p(x) \in nW\} \cap K \qquad (n = 1, 2, \ldots).$$

As $\bigcup_{n=1}^{\infty} nW = X$, it follows $\bigcup_{n=1}^{\infty} L_n = K$, and $L_n \subseteq L_{n+1}$ implies

$$\lim_n \lambda(L_n) = \lambda(K).$$

Finally we have that p is bounded on some measurable set of positive measure, hence the statement follows from Theorem 3.8.

In order to obtain the global analogue of Theorem 3.9 we note that Theorem 3.6 obviously implies

THEOREM 3.10. *Let G be a topological abelian group and let X be a locally convex topological vector space. Then any algebraic polynomial from G into X, which is continuous at a point, is continuous on the component of that point.*

THEOREM 3.11. *Let G be a locally compact abelian group and let X be a locally convex and locally bounded topological vector space. Then any measurable algebraic polynomial from G into X is continuous.*

We present a further regularity property for algebraic polynomials on topological groups, concerning a problem of Mazur. He asked about 1935 the following question ([TSC], Problem 24): Let X be a Banach space and a an additive function from X into $\mathbf{C}$ with the property, that $a \circ \varphi$ is measurable for any continuous function $\varphi : [0,1] \to X$. Is a continuous? This question was answered affirmatively in [LAB] by the following theorem.

THEOREM 3.12. *Let X be a Banach space, Y a topological vector space and $a : X \to Y$ an additive function. If $a \circ \varphi$ is measurable for any continuous function $\varphi : [0,1] \to X$, then a is continuous.*

The following more general theorem can be found in [LIP].

THEOREM 3.13. *Let G, H be topological abelian groups, G is metrizable, complete, connected and locally arcwise connected and let $a : G \to H$ be an additive function. If $a \circ \varphi$ is measurable for any continuous function $\varphi : [0,1] \to G$, then a is continuous.*

Using this theorem, here we give the following generalization.

THEOREM 3.14. *Let G be a metrizable topological abelian group which is complete, connected and locally arcwise connected, further let X be a locally convex topological vector space and let $p : G \to X$ be an algebraic polynomial.*

37

If $p \circ \varphi$ is measurable for any continuous function $\varphi : [0,1] \to G$, then p is continuous.

PROOF: Let $p = \sum_{k=0}^{n} A_k^*$ be the canonical representation of p. By Lemma 1.3 we have

$$A_n(y_1, y_2, \ldots, y_n) = \frac{1}{n!} \Delta_{y_1, y_2, \ldots, y_n} p(x) =$$

$$= \frac{1}{n!} \sum_{0 \le i_1 < i_2 < \cdots < i_k \le n} (-1)^{n-k} p(x + y_{i_1} + y_{i_2} + \cdots + y_{i_k})$$

for all $x, y_1, y_2, \ldots, y_n$ in G, which shows that the function

$$t \mapsto A_n(\varphi(t), y_2, \ldots, y_n)$$

is measurable for any continuous function $\varphi : [0,1] \to G$ and for all x, y_1, $y_2, \ldots, y_n$ in G. Using the symmetry of A_n and Theorem 3.13 we have that A_n is continuous in each variable. From the theorem of Baire [BOU2, IX.4. p.194.] it follows that A_n is continuous at at least one point. Then, using the connectedness of G, similarly as in Theorem 3.6 the continuity of A_n follows everywhere.

Further generalization of Mazur's problem will be considered in Section 5.

REFERENCES 3.15.

For the classical results and for further references see [ACZ4], [ACZ6], [ACZ7], [ACZ8], [ACZ13], [ACZ14], [ACZ15], [ACZ16], [ACZ18], [ALE], [ANG1], [ANG2], [BAK2], [BAN], [CAU], [DJO3], [FRE1], [FRE2], [FRE3], [FRE4], [GAJ1], [GAJ2], [GAJ3], [GAJ6], [GER2], [GHE1], [GHE3], [GIR1], [HAS4], [ING], [JAR1], [JAR2], [KCZ4], [KEM1], [KES], [KUR3], [LAB], [LIJ1], [LIJ2], [LIP], [MAZ], [MCK1], [MCK3], [MCK5], [OST], [PGA1], [PGM], [POP1], [POP2], [RES], [SZÉ1], [SZÉ7], [SZÉ8], [SZÉ13], [SZÉ22], [SZÉ25].

4. Exponential polynomials on semigroups

From now on we will simultaneously consider homomorphisms of a given commutative topological semigroup G into the additive group and into the multiplicative semigroup of a commutative topological ring R. To make a distinction also in the terminology, in such cases concerning additive functions and polynomials from G into R we always refer to the additive group structure of R and the homomorphisms of G into the multiplicative semigroup of R will be called exponentials from G into R. The set of all continuous nonzero exponentials from G into R equipped with the compact-open topology will be denoted by $\mathcal{E}(G, R)$. By "nonzero" we always mean "not identically" zero, but it is obvious, that if G is a group, then a nonzero exponential is never vanishing. In any case $\mathcal{E}(G, R)$ is a commutative semigroup with respect to pointwise multiplication in R.

Exponentials with values in $\mathbf{R}$ or in $\mathbf{C}$ will be called *real* or *complex exponentials*, respectively. The continuous nonzero complex exponentials on G play a very important role in our investigations. Sometimes they are called *generalized characters* of G. In particular, $\mathbf{T}$-valued continuous complex exponentials on G are the *characters* of G. The following two lemmata are of technical importance.

LEMMA 4.1. *Let G be a commutative semigroup, R a commutative ring, S an R-module, $m : G \to R$ an exponential, n a positive integer, $A : G \to S$ a symmetric n-additive function, $q : G \to S$ a polynomial of degree at most $n - 1$ and $f = m(A^* + q)$. Then, for every $y \in G$ there exists a polynomial $q_y : G \to S$ of degree at most $n - 1$ such that*

$$\Delta_y^N f(x) = m(x)\big[(m(y) - 1)^N A^*(x) + q_y(x)\big]$$

holds for all x, y in G.

PROOF: By the definition of the difference operators and by Lemma 1.2 we have

$$\Delta_y^N f(x) = \sum_{k=0}^{N} \binom{N}{k}(-1)^{N-k} f(x + ky) =$$

39

$$= \sum_{k=0}^{N} \binom{N}{k} (-1)^{N-k} m(x) m(y)^k \left[A^*(x+ky) + q(x+ky) \right] =$$

$$= \sum_{k=0}^{N} \binom{N}{k} (-1)^{N-k} m(x) m(y)^k \left[\sum_{j=0}^{n} \binom{n}{j} k^{n-j} A([x]_j, [y]_{n-j}) + q(x+ky) \right] =$$

$$= m(x) \sum_{k=0}^{N} \binom{N}{k} (-1)^{N-k} m(y)^k \left[A^*(x) + \right.$$

$$\left. + \sum_{j=0}^{n-1} \binom{n}{j} k^{n-j} A([x]_j, [y]_{n-j}) + q(x+ky) \right]$$

for all x, y in G, which yields the statement.

LEMMA 4.2. *Let G be an abelian group, K a field, X a K-linear space and $\mathcal{F}$ a translation invariant linear space of X-valued functions on G. Let n, n_i be nonnegative integers, $n \geq 1$, $m_i : G \to K$ different nonzero exponentials, $A_i : G^{n_i} \to X$ n_i-additive symmetric functions, and $q_i : G \to X$ polynomials of degree at most $n_i - 1$ $(i = 1, 2, \ldots, n)$. If the function*

$$\sum_{i=1}^{n} m_i (A_i^* + q_i)$$

belongs to $\mathcal{F}$, then there exist polynomials $r_i : G \to X$ of degree at most $n_i - 1$ such that

$$m_i (A_i^* + r_i)$$

belongs to $\mathcal{F}$ for $i = 1, 2, \ldots, n$.

PROOF: Let

$$f = \sum_{i=1}^{n} m_i (A_i^* + q_i),$$

then it follows

$$\check{m}_1 f = A_1^* + q_1 + \sum_{i=2}^{n} \check{m}_1 m_i (A_i^* + q_i).$$

Let $y \in G$ such that $m_1(y) \neq m_n(y)$ and we apply Δ_y on both sides of the above equation. Then by Lemma 4.1 we get for all x in G

$$\check{m}_1(x) \sum_{k=0}^{n_1+1} \binom{n_1+1}{k} (-1)^{n_1+k-1} \check{m}_1(ky) f(x+ky) =$$

40

$$= \sum_{i=2}^{n} \check{m}_1(x) m_i(x) \big[(\check{m}_1(y) m_i(y) - 1)^{n_1+1} A_i^*(x) + q_{i,y}(x) \big],$$

where $q_{i,y} : G \to X$ is a polynomial of degree at most $n_i - 1$ $(i = 2, \ldots, n)$. If we multiply both sides of this equation by $m_1(x)$, then by the properties of $\mathcal{F}$ it follows, that the function

$$\sum_{i=2}^{n} m_i(c_i A_i^* + q_i)$$

belongs to $\mathcal{F}$, where $c_i \in K$ with $c_n \neq 0$, and $q_i : G \to X$ is a polynomial of degree at most $n_i - 1$ $(i = 2, 3, \ldots, n)$. Continuing this argument we get the statement.

LEMMA 4.3. *Let G be an abelian group, K a field, X a K-linear space, n a positive integer, $m_i : G \to K$ a nonzero exponential, $p_i : G \to X$ a polynomial and $m_i \neq m_j$ for $i \neq j$. If $\sum_{i=1}^{n} m_i p_i = 0$, then $p_i = 0$ $(i, j = 1, 2, \ldots, n)$.*

PROOF: We can apply Lemma 4.2 with $\mathcal{F} = \{0\}$ and $p_i = A_i^* + q_i$, where A_i^* is the leading term of p_i $(i = 1, 2, \ldots, n)$. Then we have $A_i^* + q_i = 0$, which implies $A_i^* = 0$ by Theorem 2.3 $(i = 1, 2, \ldots, n)$. Then our statement follows.

Let G be a commutative topological semigroup, R a commutative topological ring and S a topological R-module. A continuous function $f : G \to S$ will be called an *exponential polynomial* from G into S, if it has a representation in the form

$$f = \sum_{i=1}^{n} m_i p_i,$$

where n is a positive integer, $m_i : G \to R$ is an exponential, and $p_i : G \to S$ is an algebraic polynomial $(i = 1, 2, \ldots, n)$. The set of all exponential polynomials from G to S will be denoted by $\mathcal{EP}(G, S)$. Obviously $\mathcal{P}(G, S) \subseteq \mathcal{EP}(G, S)$ and $\mathcal{E}(G, S) \subseteq \mathcal{EP}(G, S)$, further $\mathcal{EP}(G, S)$ is translation invariant. With the compact-open topology of $\mathcal{C}(G, S)$ the set $\mathcal{EP}(G, S)$ is a topological module over R. Similarly as in the case of polynomials, this notion depends on the topologies on G, R and S. The largest class of exponential polynomials on G is the one with respect to the discrete topology of G, which will be denoted by $\mathcal{EP}(G_d, S)$. The elements of this class will be called *algebraic exponential polynomials*. In the cases $R = S = \mathbf{R}$ or $R = S = \mathbf{C}$ we speak about *real* or *complex exponential polynomials*, respectively, and we denote $\mathcal{EP}(G, \mathbf{C})$ simply by $\mathcal{EP}(G)$.

41

The following two theorems are easy consequences of the definition and of Theorems 2.1 and 2.2.

THEOREM 4.4. *Let G be a commutative topological semigroup, K a topological field and X a topological linear space over K. Then the set of all exponential polynomials from G into X is a topological linear space over K.*

THEOREM 4.5. *Let G be a commutative topological semigroup and R a topological ring. The set of all exponential polynomials from G into R is a*
 (i) *(commutative) topological ring, if R is a (commutative) topological ring;*
(ii) *(commutative) topological algebra over the topological field K, if R is a (commutative) topological algebra over K.*

If G is an abelian group, R is a field and S is an R-linear space then, by Lemma 4.3, the given representation of f is unique, whenever the exponentials m_i are nonzero and different. Then it will be called the *canonical representation* of f. Supposing that the polynomials p_i are different from zero we call the number n the *order* of f and we define the *type* of f as $(\deg p_1, \ldots, \deg p_n)$. The order and the type of the zero exponential polynomial is defined as 0 and (0), respectively. In the case $R = \mathbf{C}$ we call f a *trigonometric polynomial*, whenever all exponentials in its canonical representation are continuous and of absolute value 1, and all polynomials are constant. Hence a trigonometric polynomial on a commutative topological group with values in a complex linear space X is a linear combination of characters of G with coefficients in X. In the case $X = C$ we call it a complex trigonometric polynomial. The set of all trigonometric polynomials from G into X will be denoted by $\mathcal{TP}(G, X)$, and $\mathcal{TP}(G)$ denotes the set of all complex trigonometric polynomials on G.

On commutative topological groups the following theorem is a generalization of Theorem 2.4.

THEOREM 4.6. *Let G be a topological abelian group, X a complex topological vector space, $k, n_1, n_2, \ldots, n_k$ nonnegative integers, $m_i : G \to \mathbf{C}$ different nonzero continuous exponentials and let $\{p_{i,\alpha}\}$ be nets of polynomials of degree at most n_i from G into X for $i = 1, 2, \ldots, k$. If*

$$f_\alpha = \sum_{i=1}^{k} m_i p_{i,\alpha},$$

42

and the net $\{f_\alpha\}$ converges uniformly on all compact set, then all the nets $\{p_{i,\alpha}\}$ $(i = 1, 2, \ldots, k)$ converge uniformly on all compact sets and

$$\lim_\alpha f_\alpha = \sum_{i=1}^k m_i \lim_\alpha p_{i,\alpha}.$$

PROOF: Let

$$p_{i,\alpha} = A_{i,\alpha}^* + q_{i,\alpha},$$

where $A_{i,\alpha}^*$ is a homogeneous polynomial of degree n_i and $q_{i,\alpha}$ is a polynomial of degree at most $n_i - 1$ $(i = 1, 2, \ldots, k)$. We have

$$\check{m}_1 f_\alpha = A_{i,\alpha}^* + q_{i,\alpha} + \sum_{i=2}^k \check{m}_1 m_i (A_{i,\alpha}^* + q_{i,\alpha}).$$

Let $y \in G$ such that $m_1(y) \neq m_n(y)$ and we apply Δ_y on both sides of the above equation. Then by Lemma 4.1 we get for all x in G

$$\check{m}_1(x) \sum_{j=0}^{n_1+1} \binom{n_1+1}{j} (-1)^{n_1+j-1} \check{m}_1(jy) f_\alpha(x + jy) =$$

$$= \sum_{i=2}^k \check{m}_1(x) m_i(x) [(\check{m}_1(y) m_i(y) - 1)^{n_1+1} A_{i,\alpha}^*(x) + q_{i,\alpha,y}(x)],$$

where $q_{i,\alpha,y} : G \to X$ is a polynomial of degree at most $n_i - 1$ $(i = 2, \ldots, k)$. If we multiply both sides of this equation by $m_1(x)$, then by induction we have that the nets of polynomials

$$\left\{ \sum_{i=2}^k m_i [(\check{m}_1(y) m_i(y) - 1)^{n_i+1} A_{i,\alpha}^* + q_{i,\alpha,y}] \right\}$$

converge uniformly on all compact sets for $i = 2, 3, \ldots, k$. As $m_1(y) \neq m_k(y)$, by Theorem 2.4 we see that $\{A_{k,\alpha}^*\}$ converges uniformly on all compact sets. Continuing this process, we get the statement.

The *normal exponential polynomials* are the exponential polynomials having a representation

$$f = \sum_{i=1}^n m_i p_i$$

where the p_i's are normal polynomials. *Exponential monomials* are defined as products of a monomial and a complex exponential. The following lemma is obvious.

LEMMA 4.7. *The set of all normal exponential polynomials on the commutative topological semigroup G is the subalgebra generated by all continuous complex homomorphisms and continuous complex exponentials of G in the algebra of all complex valued functions on G.*

In other words, normal exponential polynomials on commutative topological semigroups are just the polynomials of continuous complex homomorphisms of G into the additive group and into the multiplicative semigroup of **C**. The following lemma is a consequence - and generalization - of Lemma 2.8.

LEMMA 4.8. *Let G be a commutative topological semigroup, n a positive integer, α an n-dimensional multi-index, m a continuous nonzero complex exponential on G and let $a = (a_1, a_2, \ldots a_n)$, where $a_1, a_2, \ldots a_n$ are linearly independent continuous real additive functions on G. Then the smallest translation invariant linear space of continuous complex valued functions on G, which contains the exponential monomial ma^α has a basis consisting of all the monomials ma^β with $\beta \leq \alpha$. (Consequently, its dimension is $(\alpha_1 + 1)(\alpha_2 + 1) \ldots (\alpha_n + 1)$.)*

REFERENCES 4.9.

For further results and references concerning exponential polynomials see [ACZ7], [ANS], [BEL], [CAR1], [CAR2], [ELL1], [ELL2], [ENG], [GAJ6], [GIL1], [GIL2], [KEM4], [LAI2], [MCK2], [MCK5], [NOV], [OST], [REI1], [REI2], [STO], [SZÉ7], [SZÉ9], [SZÉ10], [SZÉ14], [SZÉ15].

5. Regularity properties of exponential polynomials

As any polynomial is an exponential polynomial, the following theorems are generalizations of those in Section 3.

THEOREM 5.1. *Let G be an abelian group and let X be a locally convex topological vector space over the complex field. Then any bounded exponential polynomial from G into X is a trigonometric polynomial.*

PROOF: We apply Lemma 4.2 by letting $\mathcal{F}$ to be the space of all bounded X-valued functions on G. Then by Lemma 4.2 and Theorem 3.1 it is enough to prove our statement for exponential polynomials of order 1, that is, for $f = mp$. If p is of degree zero, that is constant, then m is bounded. Indeed, otherwise there would exist a sequence $\{x_n\} \subseteq G$ for which $|m(x_n)| \geq n$. On the other hand, if $W \subseteq X$ is a convex balanced neighborhood of zero with $p \notin W$, then there is an $\alpha > 0$ for which $m(x_n)p \in \alpha W$ $(n = 1, 2, \dots)$. As $|nm(x_n)^{-1}| \leq 1$, hence $np = nm(x_n)^{-1}m(x_n)p \in \alpha W$. For $n > \alpha$ we have $\alpha n^{-1} < 1$ and hence $p \in \alpha n^{-1}W \subseteq W$, which is a contradiction. That is, m is bounded. If at the point $x_0 \in G$ we have $|m(x_0)| \neq 1$, then $|m(x_0)|$ or $|m(-x_0)|$ is greater than 1, and hence $m(nx_0)$ or $m(-nx_0)$ has arbitrary great absolute value, contradicting the boundedness of m.

Suppose now that we have proved our statement whenever p is of degree at most $N - 1$, and let p be of degree $N \geq 1$. Then there exists an y in G such that $p(x + y) \neq p(x)$ for all x, that is, $\Delta_y p$ is not identically zero, and it is a polynomial of degree at most $N - 1$. Further for $x \in G$ we have

$$m(x)\Delta_y p(x) = m(-y)m(x + y)p(x + y) - m(x)p(x),$$

which implies that $m\Delta_y p$ is bounded. Hence by induction on the degree of p it follows $|m| = 1$, which implies that p is also bounded. Then p is constant by Theorem 3.1, hence our theorem is proved.

The next lemma is a local version of Lemma 4.2. We note, that if G is a topological abelian group and X is a linear space, then a linear space $\mathcal{F}$ of X-

45

valued functions on G is called *locally translation invariant*, if for any f in $\mathcal{F}$ there exists a neighborhood $U \subseteq G$ of zero such that $\tau_y f \in \mathcal{F}$ for all y in U. Here and everywhere $\tau_y f$ denotes the *translate* of f by y : $\tau_y f(x) = f(x + y)$ for all x, y in G.

LEMMA 5.2. *Let G be a topological abelian group which is generated by any neighborhood of zero, X a complex linear space and let $\mathcal{F}$ be a locally translation invariant vector space of X - valued functions on G. Let n, n_i be nonnegative integers, $n \geq 1$, $m_i : G \to \mathbb{C}$ different nonzero exponentials, $A_i : G^i \to X$ n_i- additive symmetric functions, and let $q_i : G \to X$ be algebraic polynomials of degree at most $n_i - 1$ $(i = 1, 2, \ldots, n)$. If the function*

$$f = \sum_{i=1}^{n} m_i(A_i^* + q_i)$$

belongs to $\mathcal{F}$, then there exist algebraic polynomials $r_i : G \to X$ of degree at most $n_i - 1$ such that

$$m_i(A_i^* + r_i)$$

belongs to $\mathcal{F}$ for $i = 1, 2, \ldots, n$.

PROOF: Let $U \subseteq G$ be a neighborhood of zero such that $\tau_y f \in \mathcal{F}$ for all y in U, and let $V \subseteq U$ such that $ky \in U$ whenever $y \in V$ for $k = 0, 1, \ldots, n_1$, where n_1 denotes the degree of q_1. Then it follows

$$\check{m}_1 f = A_1^* + q_1 + \sum_{i=2}^{n} \check{m}_1 m_i (A_i^* + q_i).$$

As V generates G, there exists a $y \in V$ such that $m_1(y) \neq m_n(y)$ and we apply Δ_y on both sides of the above equation. Then by Lemma 4.1 we get for all x in G

$$\check{m}_1(x) \sum_{k=0}^{n_1+1} \binom{n_1 + 1}{k} (-1)^{n_1+k-1} \check{m}_1(ky) f(x + ky) =$$

$$= \sum_{i=2}^{n} \check{m}_1(x) m_i(x) \left[(\check{m}_1(y) m_i(y) - 1)^{n_1+1} A_i^*(x) + q_{i,y}(x) \right],$$

where $q_{i,y} : G \to X$ is an algebraic polynomial of degree at most $n_i - 1$ $(i = 2, \ldots, n)$. If we multiply both sides of this equation by $m_1(x)$, then, by the properties of $\mathcal{F}$ it follows, that the function

$$\sum_{i=2}^{n} m_i(c_i A_i^* + r_i)$$

46

belongs to $\mathcal{F}$, where $c_i \in \mathbf{C}$ with $c_n \neq 0$, and $r_i : G \to X$ is an algebraic polynomial of degree at most $n_i - 1$ $(i = 2, 3, \ldots, n)$. Continuing this argument we get the statement.

THEOREM 5.3. *Let G be a topological abelian group which is generated by any neighborhood of zero and let X be a linear space. If an algebraic exponential polynomial from G into X vanishes on a nonvoid open set, then it vanishes everywhere.*

PROOF: Obviously we may suppose, that the exponential polynomial vanishes on a neighborhood of zero. If $\mathcal{F}$ denotes the set of all X-valued functions on G which vanish on a neighborhood of zero, then $\mathcal{F}$ is a locally translation invariant vector space and we can apply Lemma 5.2, hence it is enough to prove our theorem for algebraic exponential polynomials of order 1. But this is a consequence of Theorem 3.2.

THEOREM 5.4. *Let G be a locally compact abelian group which is generated by any neighborhood of zero and let X be a linear space. If an algebraic exponential polynomial from G into X vanishes on a measurable set of positive measure, then it vanishes everywhere.*

PROOF: Let the algebraic exponential polynomial $f : G \to X$ have the canonical representation

$$f = \sum_{i=1}^{n} m_i(A_i^* + q_i),$$

where the $A_i : G^{n_i} \to X$ are n_i-additive symmetric functions, and $q_i : G \to X$ are algebraic polynomials of degree at most $n_i - 1$ $(i = 1, 2, \ldots, n)$. We prove the statement by induction on n, and for $n = 1$ it follows from Theorem 3.3. Suppose that it has been proved for all values less than n and now we prove it for n. We assume that f vanishes on the compact set $K \subseteq G$ of positive measure. It follows, that

$$\check{m}_1 f = A_1^* + q_1 + \sum_{i=2}^{n} \check{m}_1 m_i(A_i^* + q_i)$$

vanishes on K, and let $V \subseteq G$ be a neighborhood of zero for which $\lambda(K \cap K - y \cap \cdots \cap K - (n_1 + 1)y) > 0$ whenever y is in V. By the same argument as in Lemma 5.2 there exists an element y in V for which $m_1(y) \neq m_2(y)$. Let

$$K_1 = K \cap K - y \cap \cdots \cap K - (n_1 + 1)y,$$

47

then $x + ky$ belongs to K for all x in K_1 $(k = 0, 1, \ldots, n_1 + 1)$ and we apply Δ_y on both sides of the above equation. Then by Lemma 4.1 we get for all x in G

$$\check{m}_1(x) \sum_{k=0}^{n_1+1} \binom{n_1 + 1}{k} (-1)^{n_1+k-1} \check{m}_1(ky) f(x + ky) =$$

$$= \sum_{i=2}^{n} \check{m}_1(x) m_i(x) \big[(\check{m}_1(y) m_i(y) - 1)^{n_1+1} A_i^*(x) + q_{i,y}(x) \big],$$

where $q_{i,y} : G \to X$ is an algebraic polynomial of degree at most $n_i - 1$ $(i = 2, \ldots, n)$. If we multiply both sides of this equation by $m_1(x)$, then we have

$$\sum_{i=2}^{n} m_i(x) \big[(\check{m}_1(y) m_i(y) - 1)^{n_1+1} A_i^*(x) + q_{i,y}(x) \big] = 0$$

for all x in K_1 and $\lambda(K_1) > 0$ implies by induction

$$(\check{m}_1(y) m_i(y) - 1)^{n_1+1} A_i^*(x) + q_{i,y}(x) = 0$$

for all x in G. Here $\check{m}_1(y) m_2(y) \neq 1$ and the degree of $q_{2,y}$ is at most $n_2 - 1$, hence $A_2^* = 0$ by Theorem 2.3. Continuing this argument we get our statement.

LEMMA 5.5. *Let G be a topological abelian group and let X be a locally convex topological vector space over the complex field. Then all algebraic polynomials and exponentials in the canonical representations of exponential polynomials from G to X are continuous.*

PROOF: We apply Lemma 4.2 by letting $\mathcal{F}$ to be the set of all continuous X-valued functions on G. Then by Lemma 4.2 and Theorem 3.4 it is enough to prove our statement for exponential polynomials of order 1, that is, for $f = mp$. If p is of degree 0, that is constant, and we suppose that m is not continuous at zero, then, by $m(0) = 1$ we find an $\epsilon > 0$ and a sequence $\{x_n\} \subseteq G$ with $x_n \to 0$ and $|m(x_n) - 1| \geq \epsilon$. Let $W \subseteq X$ be a convex balanced neighborhood of zero with $p \notin W$. As the function $x \mapsto m(x)p$ is continuous, hence $m(x_n)p - p \in \epsilon W$ whenever n is large enough. On the other hand, $\epsilon |m(x_n) - 1|^{-1} \leq 1$ hence $\epsilon p \in \epsilon W$ and $p \in W$, which is impossible. That is, m is continuous at zero, and by Theorem 3.6 it follows its continuity everywhere. The remaining parts of the lemma can be proved as in Theorem 5.1.

THEOREM 5.6. *Let G be a topological abelian group which is generated by any neighborhood of zero and let X be a locally convex topological vector space*

48

over the complex field. Then any algebraic exponential polynomial from G into X, which is continuous on some nonvoid open set, is continuous everywhere.

PROOF: We may suppose that the exponential polynomial is continuous on a neighborhood of zero. If $\mathcal{F}$ denotes the set of all X-valued functions on G which are continuous on a neighborhood of zero, then $\mathcal{F}$ is a locally translation invariant vector space and we can apply Lemma 5.2, hence by Lemma 5.5 and Theorem 3.5 it is enough to prove our theorem for algebraic exponential polynomials of order 1, that is for $f = mp$, which is continuous on a neighborhood $U \subseteq G$ of zero. First let p be a nonzero constant. Then m is continuous on U and Theorem 3.5 implies that it is continuous everywhere. Suppose that we have proved the continuity of m and p for all p of degree at most $N - 1$, and let the degree of p be $N \geq 1$. Let $V \subseteq G$ be a neighborhood of zero with $V + V \subseteq U$. As G is generated by V, there exists a y in V for which $\Delta_y p$ is not identically zero on V. On the other hand, the degree of $\Delta_y p$ is at most $N - 1$ and in the equation

$$m(x)\Delta_y p(x) = m(-y)f(x + y) - f(x)$$

the right hand side is continuous on V, hence by induction we have the continuity of m. Then obviously p is continuous on U, and hence on G, by Theorem 3.6, which proves our theorem.

THEOREM 5.7. *Let G be a topological abelian group which is generated by any neighborhood of zero and let X be a locally convex topological vector space over the complex field. Then any algebraic exponential polynomial $f : G \to X$, which is bounded on a nonvoid open set, can be expressed in the form*

$$f = \sum_{i=1}^{n} m_i \gamma_i p_i,$$

where $\gamma_i : G \to X$ is an exponential with $|\gamma_i| = 1$, $m_i : G \to \mathbf{C}$ is a continuous exponential and $p_i : G \to X$ is a polynomial $(i = 1, 2, \ldots, n)$.

PROOF: We may suppose that f is bounded on a neighborhood of zero. If $\mathcal{F}$ denotes the set of all X-valued functions on G which are bounded on a neighborhood of zero, then $\mathcal{F}$ is a locally translation invariant vector space and we can apply Lemma 5.2, hence it is enough to prove our theorem for algebraic exponential polynomials of order 1. Let $f = mp$ and suppose that it is bounded on U. First let p be a nonzero constant and let $W \subseteq X$ be a

49

convex, balanced and absorbing neighborhood of zero for which $p \notin W$. Let $\alpha > 0$ be such that $m(x)p \in \alpha W$ for all x in G. If m is unbounded on U, then there exists an $x \in U$ with $|m(x)| > \alpha$ and hence $\alpha m(x)^{-1}W \subseteq W$, further $p \in \alpha m(x)^{-1}W \subseteq W$, which is a contradiction. Let $\mu = |m|$ and $\gamma = m|m|^{-1}$. Obviously $|\gamma| = 1$ and $\mu > 0$ further $\gamma, \mu : G \to \mathbf{C}$ are generalized exponentials. As $A = \ln \mu$ is additive and bounded on U, it follows from Theorem 3.5 that it is continuous. Hence $m = \gamma \exp A$ is the desired representation. Our theorem is proved.

THEOREM 5.8. *Let G be a locally compact abelian group which is generated by any neighborhood of zero and let X be a locally convex and locally bounded topological vector space over the complex field. Then any algebraic exponential polynomial from G into X, which is measurable on a measurable set of positive measure, is continuous.*

PROOF: We use the notations and the method of Theorem 5.4. Using the local boundedness of X we may assume that f is bounded on the compact set $K \subseteq G$ with positive measure $\lambda(K) > 0$. First let $n = 1$ and $f = mp$ where p is a nonzero constant. By the same method as above we obtain that m is bounded on K, and hence on $K + K$ too, but this latter set has a nonvoid interior by the Steinhaus theorem ([KUR3], [HAL]). By Theorem 5.7 we have $m = \mu \gamma$, where $\mu, \gamma : G \to \mathbf{C}$ are exponentials, μ is continuous and $|\gamma| = 1$. It follows that γ is measurable on K. Let $\epsilon > 0$ be arbitrary. Then there exists a neighborhood $U \subseteq G$ of zero such that

$$\int_K |\gamma(x+y) - \gamma(y)| d\lambda(y) < \epsilon \lambda(K)$$

holds for all x in U. On the other hand

$$|\gamma(x) - 1| = \frac{1}{\lambda(K)} \int_K |\gamma(x) - 1| d\lambda(y) = \frac{1}{\lambda(K)} \int_K |\gamma(x+y) - \gamma(y)| d\lambda(y) < \epsilon$$

whenever $x \in U$, that is, γ is continuous at zero. This implies the continuity of γ and of m everywhere.

Now let p be of degree $N \geq 1$ and suppose that our statement has been proved for $n = 1$ and $\deg p \leq N - 1$. Let $U \subseteq G$ be a neighborhood of zero for which $y \in U$ implies $\lambda(K \cap K - y) > 0$. Let $y \in U$ be an element for which $\Delta_y p$ is not constant. Then by the identity

$$m(x)\Delta_y p(x) = m(-y)f(x+y) - f(x)$$

50

we see that $m\triangle_y p$ is measurable on the set $K \cap K - y$ of positive measure, and hence it follows our statement by induction (with respect to the degree of p) for $n = 1$. From this step the proof continues like in Theorem 5.4.

The following theorem can be proved similarly as above.

THEOREM 5.9. *Let G be a locally compact abelian group which is generated by any neighborhood of zero and let X be a locally convex and locally bounded topological vector space over the complex field. Then any algebraic exponential polynomial $f : G \to X$, which is bounded on a measurable set of positive measure, can be expressed in the form*

$$f = \sum_{i=1}^{n} m_i \gamma_i p_i,$$

where $\gamma_i : G \to X$ is an exponential with $|\gamma_i| = 1$, $m_i : G \to \mathbf{C}$ is a continuous exponential and $p_i : G \to X$ is a polynomial $(i = 1, 2, \ldots, n)$.

Theorem 5.8 implies its global version by the observation that the component of the zero is a closed connected subgroup.

THEOREM 5.10. *Let G be a locally compact abelian group and X a locally convex and locally bounded topological vector space over the complex field. Then any measurable algebraic exponential polynomial from G into X is continuous.*

Concerning Mazur's problem (see Section 3) we prove the following generalization of Theorem 3.14.

THEOREM 5.11. *Let G be a metrizable topological abelian group which is complete, connected and locally arcwise connected, further let X be a metrizable locally convex topological vector space and $f : G \to X$ an algebraic exponential polynomial. If $f \circ \varphi$ is measurable for any continuous function $\varphi : [0, 1] \to G$, then f is continuous.*

PROOF: We apply Lemma 4.2 by letting $\mathcal{F}$ to be the space of all complex functions on G, which are measurable along any continuous path in G. Then by Lemma 4.2 and Theorem 3.1 it is enough to prove our statement for algebraic exponential polynomials of order 1, that is, for $f = mp$. If p is of degree zero, that is constant, then it follows, that $m \circ \varphi$ is measurable for any continuous function $\varphi : [0,1] \to G$, hence by Theorem 3.13 m is continuous, that is, f is continuous. If p is nonconstant, then we can apply induction with respect to its degree, and our statement follows in the same way as in Theorem 5.1.

REFERENCES 5.12.

A systematic study of regularity properties of polynomials and exponential polynomials on topological abelian groups can be found in [SZÉ13], [SZÉ14]. In [JAR2] fundamental theorems on the regularity of solutions of general functional equations are given. For further results and references see [ACZ6], [ACZ7], [ACZ11], [ACZ13], [ACZ16], [ACZ18], [BAK2], [BEL], [CAR1], [CAR2], [GAJ6], [JAR1], [KCZ4], [KEM4], [KUR3], [LIP], [MCK2], [MCK5], [PGA1], [PGM], [SZÉ7], [SZÉ10].

6. Exponential polynomials on special abelian semigroups

In this section we collect some well-known theorems about polynomials and exponential polynomials on special abelian semigroups and groups.

THEOREM 6.1. *Any polynomial on a compact abelian semigroup with values in a locally convex topological vector space is constant.*

PROOF: This follows from Theorem 3.1, as any compact set in a locally convex topological vector space is bounded.

THEOREM 6.2. *Any polynomial on a torsion group with values in a torsion-free abelian semigroup is constant.*

PROOF: It is enough to show that any additive function is zero. Indeed, as any element x of a torsion group has a finite order, for any x there exists a positive integer n with $nx = 0$. Then, for any additive function a it follows $0 = a(0) = a(nx) = na(x)$, which implies our statement.

THEOREM 6.3. *Let $G = \mathbf{N}, \mathbf{Z}$ or $\mathbf{Q}$. Any additive function $a : G \to \mathbf{C}$ has the form*

$$a(x) = cx \qquad (x \in G)$$

with some complex constant c. Any n-additive function $A : G^n \to \mathbf{C}$ is symmetric and has the form

$$A(x_1, x_2, \ldots, x_n) = c x_1 x_2 \ldots x_n \qquad (x_1, x_2, \ldots, x_n \in G)$$

with some complex constant c. Any polynomial $p : G \to \mathbf{C}$ of degree at most n has the form

$$p(x) = \sum_{k=0}^{n} c_k x^k \qquad (x \in G)$$

with some complex constants c_k $(k = 0, 1, \ldots, n)$.

PROOF: The additivity of $a : G \to \mathbf{C}$ implies $a(nx) = na(x)$ for any x in G and for any integer n, hence the first statement is true for $G = \mathbf{N}$ or $\mathbf{Z}$.

In order to obtain the first statement for $G = \mathbf{Q}$ we set first $x = 1/n$, then $x = 1/m$. The rest of the theorem is then an easy consequence.

We recall that a Hamel-basis of $\mathbf{R}$ is a basis of $\mathbf{R}$ considered as a linear space over $\mathbf{Q}$ (see [HAM]).

THEOREM 6.4. *Let H be a Hamel-basis of $\mathbf{R}$. Any additive function $a : \mathbf{R}_d \to \mathbf{C}$ has the form*

$$a(x) = \sum_{h \in H} \mathbf{p}_h(x) a_0(h) \qquad (x \in \mathbf{R})$$

with some function $a_0 : H \to \mathbf{C}$, where $\mathbf{p}_h : \mathbf{R} \to \mathbf{Q}$ is the projection corresponding to $h \in H$. Any n-additive function $A : \mathbf{R}_d^n \to \mathbf{C}$ has the form

$$A(x_1, x_2, \ldots, x_n) = \sum_{h_1, \ldots, h_n \in H} p_{h_1}(x_1) p_{h_2}(x_2) \ldots p_{h_n}(x_n) A_0(x_1, x_2, \ldots, x_n)$$

with some function $A_0 : H^n \to \mathbf{C}$. The function A is symmetric if and only if A_0 is symmetric.

PROOF: The theorem is a consequence of the fact that by Theorem 6.1 any complex additive function is a $\mathbf{Q}$-linear functional on the $\mathbf{Q}$-linear space $\mathbf{R}_d$.

THEOREM 6.5. *Any continuous additive function $a : \mathbf{R} \to \mathbf{C}$ has the form*

$$a(x) = cx \qquad (x \in \mathbf{R})$$

with some complex constant c. Any continuous n - additive function $A : \mathbf{R}^n \to \mathbf{C}$ is symmetric and has the form

$$A(x_1, x_2, \ldots, x_n) = c x_1 x_2 \ldots x_n \qquad (x_1, x_2, \ldots, x_n \in \mathbf{R})$$

with some complex constant c. Any polynomial $p : \mathbf{R} \to \mathbf{C}$ of degree at most n has the form

$$p(x) = \sum_{k=0}^{n} c_k x^k \qquad (x \in \mathbf{R})$$

with some complex constants c_k $(k = 0, 1, \ldots, n)$.

PROOF: Let

$$F(x) = \int_0^x a(t) dt \qquad (x \in \mathbf{R}),$$

then F is differentiable and we have for all x, y in $\mathbf{R}$

$$F(x) = \int_0^x a(t)dt = \int_0^x \big[a(t+y) - a(y)\big]dt = \int_0^x a(t+y)dt - xa(y) =$$

$$= \int_y^{x+y} a(t)dt - xa(y) = F(x+y) - F(y) - xa(y),$$

which shows that a is differentiable. The additive property of a implies then that the derivative of a is constant, hence by $a(0) = 0$ the statement follows. The other statements are easy consequences.

THEOREM 6.6. *Any continuous additive function* $a : \mathbf{C} \to \mathbf{C}$ *has the form*

$$a(z) = cz + d\bar{z} \qquad (z \in \mathbf{C})$$

with some complex constants c, d. *Any polynomial* $p : \mathbf{C} \to \mathbf{C}$ *of degree at most* n *has the form*

$$p(z) = \sum_{0 \le k+l \le n} c_{k,l} z^k \bar{z}^l \qquad (z \in \mathbf{C})$$

with some complex constants $c_{k,l}$ $(k, l = 0, 1, \ldots, n)$.

PROOF: Let for $x \in \mathbf{R}$

$$f(x) = a(x), \qquad g(x) = a(ix),$$

then obviously $f, g : \mathbf{R} \to \mathbf{C}$ are continuous additive functions. Then the equation

$$a(z) = a(\Re z + i\Im z) = a(\Re z) + a(i\Im z) = f(\Re z) + g(\Im z) \qquad (z \in \mathbf{C})$$

together with Theorem 6.5 gives our first statement. The second statement is an obvious consequence.

THEOREM 6.7. *Let* G *be a commutative topological group. Any continuous nonzero complex exponential* $m : G \to \mathbf{C}$ *has the form*

$$m = \mu \exp a,$$

55

where $\mu : G \to \mathbf{C}$ is a character and $a : G \to \mathbf{R}$ is a continuous additive function.

PROOF: The identity

$$m(x) = m(x_0)m(x - x_0),$$

which holds for any exponential, shows that a nonzero exponential on a group is never vanishing. Hence $x \mapsto |m(x)|$ is a positive real exponential, $x \mapsto m(x)|m(x)|^{-1}$ is a character.

THEOREM 6.8. *Any continuous nonzero complex exponential on a compact topological group is a character.*

PROOF: This follows from Theorem 6.1 and Theorem 6.8.

Theorem 6.7 shows that the knowledge of the character group and of the group of all continuous real additive functions on a topological abelian group result a complete description of all continuous complex exponentials. For the character groups of special abelian groups see e.g. [HEW].

THEOREM 6.9. *Any character $\gamma : \mathbf{R} \to \mathbf{T}$ has the form*

$$\gamma(x) = \exp i\lambda x \qquad (x \in \mathbf{R})$$

with some real constant λ. Any continuous complex exponential $m : \mathbf{R} \to \mathbf{C}$ has the form

$$m(x) = \exp \lambda x \qquad (x \in \mathbf{R})$$

with some complex constant λ. Any exponential polynomial $f : \mathbf{R} \to \mathbf{C}$ is normal and has the form

$$f(x) = \sum_{k=1}^{n} \sum_{j=0}^{n_k} c_{k,j} x^j \exp \lambda_k x \qquad (x \in \mathbf{R})$$

with some nonnegative integers n_k and some complex constants $c_{k,j}, \lambda_k$ ($k = 1, 2, \ldots, n; j = 0, 1, \ldots, n_k$).

PROOF: By Theorem 6.5 and Theorem 6.7 it is enough to prove the first statement. If $\gamma : \mathbf{R} \to \mathbf{T}$ is any character, we let

$$F(x) = \int_0^x \gamma(t)dt \qquad (x \in \mathbf{R}),$$

56

then F is differentiable, and it is not identically zero, as its derivative is just γ. On the other hand, we have for all x, y in $\mathbf{R}$

$$\gamma(y)F(x) = \gamma(y)\int_0^x \gamma(t)dt = \int_0^x \gamma(t+y)dt = \int_y^{x+y} \gamma(t)dt = F(x+y) - F(y),$$

which shows that γ is differentiable. The exponential property of γ implies for all x in $\mathbf{R}$

$$\gamma'(x) = \gamma'(0)\gamma(x),$$

and hence the derivative of the function $x \mapsto \gamma(x)\exp(-\gamma'(0)x)$ vanishes, which yields our statement.

THEOREM 6.10. *Let n be a positive integer and let G_i be a commutative topological group for $i = 1, 2, \ldots, n$. Any continuous complex additive function $a : G_1 \times G_2 \times \cdots \times G_n \to \mathbf{C}$ has the form*

$$a(x_1, x_2, \ldots, x_n) = a_1(x_1) + a_2(x_2) + \cdots + a_n(x_n)$$

for all $(x_1, x_2, \ldots, x_n)$ in $G_1 \times G_2 \times \cdots \times G_n$, where $a_i : G_i \to \mathbf{C}$ is a continuous additive function for $i = 1, 2, \ldots, n$.

PROOF: Observe, that

$$x_i \mapsto a(0, 0, \ldots, 0, x_i, 0, \ldots, 0)$$

is continuous and additive on G_i for any $i = 1, 2, \ldots, n$, and let

$$a_i(x) = a(0, 0, \ldots, 0, x, 0, \ldots, 0)$$

where x is arbitrary in G_i.

THEOREM 6.11. *Let n be a positive integer and let G_i be a commutative topological group for $i = 1, 2, \ldots, n$. Any continuous complex exponential $m : G_1 \times G_2 \times \cdots \times G_n \to \mathbf{C}$ has the form*

$$m(x_1, x_2, \ldots, x_n) = m_1(x_1)m_2(x_2)\ldots m_n(x_n)$$

for all $(x_1, x_2, \ldots, x_n)$ in $G_1 \times G_2 \times \cdots \times G_n$, where $m_i : G_i \to \mathbf{C}$ is a continuous exponential for $i = 1, 2, \ldots, n$.

PROOF: Observe, that

$$x_i \mapsto m(0, 0, \ldots, 0, x_i, 0, \ldots, 0)$$

is a continuous exponential on G_i for any $i = 1, 2, \ldots, n$, and let

$$m_i(x) = m(0, 0, \ldots, 0, x, 0, \ldots, 0)$$

where x is arbitrary in G_i.

REFERENCES 6.12.

For further references concerning the results collected in this section see [ACZ7], [ACZ18], [HEW], [KCZ4].

58

CHAPTER 2

FOURIER-TRANSFORMATION AND MEAN PERIODIC FUNCTIONS

7. The Fourier-transform of exponential polynomials

In this section we introduce the Fourier-transform of exponential polynomials. This is a useful tool to determine exponential polynomial solutions of linear functional equations. Let G be a topological abelian group and X a complex topological vector space. As in Section 4, $\mathcal{EP}(G, X)$ denotes the set of all X-valued exponential polynomials on G. By Lemma 4.3 any f in $\mathcal{EP}(G, X)$ has a unique representation in the form

$$f = p_0 + \sum_{i=1}^{n} m_i p_i,$$

where $m_i : G \to \mathbf{C}$ is a nonzero exponential, $m_i \neq m_j$ for $i \neq j$ and $m_i \neq 1$ $(i, j = 1, 2, \ldots, n)$, further $p_i : G \to X$ is a polynomial $(i = 0, 1, \ldots, n)$. We introduce a polynomial-valued operator M on $\mathcal{EP}(G, X)$ by the formula

$$M(f) = p_0.$$

The following lemma summarizes the basic properties of M.

LEMMA 7.1. *The operator M is linear and it has the following properties:*
(i) $M(p) = p$
(ii) $M(pf) = pM(f)$
(iii) $M(\tau_y f) = \tau_y M(f)$
(iv) $M(\check{f}) = \big(M(f)\big)\check{}$
for any exponential polynomial $f : G \to X$, polynomial $p : G \to X$ and $y \in G$.

The statements are direct consequences of the definition. We note that the following property of M follows immediately: for any exponential $m : G \to \mathbf{C}$ and $c \in X$ we have

$$M(cm) = \begin{cases} 0 \ for \ m \neq 1 \\ c \ for \ m = 1. \end{cases}$$

For any exponential polynomial $f : G \to X$ we define the polynomial-valued function $f : \mathcal{E}(G) \to \mathcal{P}(G, X)$ by

$$\hat{f}(m) = M(f\check{m})$$

for any nonzero exponential $m : G \to \mathbf{C}$. Obviously, here $f\check{m}$ is an exponential polynomial from G into X and we can realize $\hat{f}$ as the polynomial coefficient of the exponential m in the canonical representation of f. For complex trigonometric polynomials this coincides with the usual Fourier-transform of f, hence in general we call it the *Fourier-transform* of the exponential polynomial f.

LEMMA 7.2. *The map $f \mapsto \hat{f}$ defined above is linear and has the following properties:*
  (i) $\hat{p}(m) = 0$ for $m \neq 1$, $\hat{p}(1) = p$;
  (ii) $(pf)\check{} = p\hat{f}(m)$;
  (iii) $(\tau_y f)\check{}(m) = m(y)(\tau_y \hat{f})(m)$;
  (iv) $(\check{f})\check{}(m) = [\hat{f}(\check{m})]\check{}$;
for any exponential polynomial $f : G \to X$, polynomial $p : G \to X$, exponential $m : G \to \mathbf{C}$ and y in G.

The statements are direct consequences of Lemma 7.1 and the definition. Also the following "inversion theorem" is obvious.

THEOREM 7.3. *Let G be a topological abelian group, X a complex topological vector space and $f : G \to X$ an exponential polynomial. Then we have*

$$f = \sum \hat{f}(m)m,$$

where the sum is taken over all nonzero complex exponentials m on G.

REFERENCES 7.4.

For further references see [SZÉ16], [SZÉ17], [SZÉ18].

8. Mean periodic functions on abelian groups

In this section we collect some basic facts from the theory of mean periodic functions which serve as fundamental tools for our later investigations.

Let G be a topological abelian group and let $\mathcal{C}(G)$ denote the set of all continuous complex valued functions on G. Equipped with the pointwise operations and with the topology of uniform convergence on compact sets, $\mathcal{C}(G)$ is a locally convex topological vector space. The dual of $\mathcal{C}(G)$ can be identified with the space $\mathcal{M}_c(G)$ of all Radon measures with compact support. The set of all continuous homomorphisms of G into the multiplicative group of nonzero complex numbers, that is, the set of all nonzero complex exponentials on G is a topological abelian group with respect to pointwise multiplication and to the topology of $\mathcal{C}(G)$. This group - sometimes called the *generalized character group* of G [ELL2] - will be denoted by $\tilde{G}$. The character group $\hat{G}$ of G is obviously a closed subgroup of $\tilde{G}$. If G is compact, then $\hat{G} = \tilde{G}$, by Theorem 6.8.

A closed translation invariant subspace of $\mathcal{C}(G)$ is called a *variety*.

LEMMA 8.1. *If $V \subseteq \mathcal{C}(G)$ is a closed subspace and*

$$V^\perp = \{\mu \in \mathcal{M}_c(G) : \mu(f) = 0 \ \text{for all} \ f \ \text{in} \ V\},$$

then $V^\perp$ is a closed ideal in $\mathcal{M}_c(G)$ if and only if V is a variety. (Then it is called the annihilator ideal of V.)

PROOF: First suppose that V is a variety. It is obvious that $V^\perp$ is a closed subspace of $\mathcal{M}_c(G)$. If $\mu \in V^\perp$, $\nu \in \mathcal{M}_c(G)$ and $f \in V$, then $\tau_y f \in V$ for all y in G, hence $\mu(\tau_y f) = 0$, which implies

$$\mu * \nu(f) = \int f d(\mu * \nu) = \int \int f(x + y) d\mu(x) d\nu(y) = 0$$

that is, $\mu * \nu \in V^\perp$. Conversely, if $V^\perp$ is a closed ideal in $\mathcal{M}_c(G)$, then for any $\mu \in V^\perp$, $f \in V$ and $\nu \in \mathcal{M}_c(G)$ we have

$$0 = \int f d(\mu * \nu) = \int \left[\int f(x + y) d\mu(x) \right] d\nu(y)$$

that is, the function $y \mapsto \mu(\tau_y f)$ annihilates $\mathcal{M}_c(G)$. This means that $\mu(\tau_y f)$ vanishes, and by the Hahn-Banach theorem $\tau_y f$ is in V.

For any f in $\mathcal{C}(G)$, the closed subspace spanned by the set of all translates of f is evidently a variety and is denoted by $\tau(f)$. For any μ in $\mathcal{M}(G)$ the set $V(\mu)$ of all functions f in $\mathcal{C}(G)$ with the property $f * \mu = 0$ is also a variety.

The set of all nonzero exponentials contained in a variety V is called the *spectrum* of V and is denoted by $sp(V)$. We call $sp(\tau(f))$ the spectrum of f and we denote it simple by $sp(f)$.

The union of all varieties of the form $V(\mu)$ with $\mu \neq 0$, endowed with the inductive limit of the topologies of the spaces $V(\mu)$, is a topological space which we denote by $\mathcal{MP}(G)$. The elements of $\mathcal{MP}(G)$ are called *mean periodic* functions. Hence a continuous complex valued function f on G is mean periodic if and only if there exists a nonzero Radon measure μ with compact support such that $f * \mu = 0$. Equivalently, this means $\tau(f) \neq \mathcal{C}(G)$. Convergence of a net in $\mathcal{MP}(G)$ means that this net is contained in the annihilator subspace of some nonzero compactly supported Radon measure and it converges uniformly on compact sets.

If the varieties $V(\mu)$ for $\mu \in \mathcal{M}_c(G)$, $\mu \neq 0$ form an inductive set, that is, for any $\mu_i \in \mathcal{M}_c(G)$, $\mu_i \neq 0$ $(i = 1, 2)$ there exists a $\mu \in \mathcal{M}_c(G)$, $\mu \neq 0$ with $V(\mu_1) \cup V(\mu_2) \subseteq V(\mu)$, then $\mathcal{MP}(G)$ is a locally convex topological vector space. This holds, for instance, if $\mu_1 * \mu_2 \neq 0$, that is, if the convolution of nonzero compactly supported measures is nonzero, which is obviously the case, if e.g. $G = \mathbf{R}$.

THEOREM 8.2. *If G is a topological abelian group for which $\mathcal{E}(G)$ is infinite, then any complex exponential polynomial on G is mean periodic.*

PROOF: Let the exponential polynomial $f : G \to \mathbf{C}$ have the canonical representation

$$f = \sum_{i=1}^{N} m_i p_i$$

and let n_i denote the degree of p_i $(i = 1, 2, \ldots, N)$. It follows

$$\check{m}_1 f = p_1 + \sum_{i=2}^{N} \check{m}_1 m_i p_i$$

and, by applying the difference operator $\Delta_y^{n_1+1}$ on both sides, we have

$$\check{m}_1(x) \sum_{k_1=0}^{n_1+1} \binom{n_1+1}{k_1} (-1)^{n_1+1-k_1} \check{m}_1(y_1)^{k_1} f(x + k_1 y) =$$

$$= \sum_{i=2}^{N} \breve{m}_1(x) m_i(x) \sum_{k_1=0}^{n_1+1} \binom{n_1+1}{k_1} (-1)^{n_1+1-k_1} \breve{m}_1(y_1)^{k_1} m_i(y_1)^{k_1} p_i(x+k_1 y) =$$

$$= \sum_{k_1=0}^{n_1+1} \breve{m}_1(x) \binom{n_1+1}{k_1} (-1)^{n_1+1-k_1} \sum_{i=2}^{N} m_i(x) \breve{m}_1(y_1)^{k_1} m_i(y_1)^{k_1} \left[\tau_{k_1 y} p_i(x) \right]$$

by Lemma 1.4. Here $m_i(y_1)^{k_1} \left[\tau_{k_1 y} p_i \right]$ is a polynomial of degree n_2, hence we can repeat our argument. Continuing this procedure, we get

$$\sum_{k_1=0}^{n_1+1} \cdots \sum_{k_N=0}^{n_N+1} \binom{n_1+1}{k_1} \cdots \binom{n_N+1}{k_N} (-1)^{n_1+\cdots+n_N+N-k_1-\cdots-k_N} \times$$

$$\times \breve{m}_1(y_1)^{k_1} \breve{m}_2(y_2)^{k_2} \ldots \breve{m}_N(y_N)^{k_N} f(x+k_1 y_1+k_2 y_2+\cdots+k_N y_N) = 0$$

for all $x, y_1, y_2, \ldots, y_N$ in G. Obviously, any element of $\tau(f)$ satisfies this equation. To show that $\tau(f) \neq \mathcal{C}(G)$, let m be any nonzero continuous complex exponential, which is different from m_i for all i. Substituting m for f into the above equation we get

$$m(x) \big(m(y_1) \breve{m}_1(y_1) - 1 \big)^{n_1+1} \ldots \big(m(y_N) \breve{m}_N(y_N) - 1 \big)^{n_N+1} = 0$$

which is impossible for all $x, y_1, y_2, \ldots, y_N$. Hence f is mean periodic.

The above proof shows that no exponential different from the m_i's can belong to the spectrum of f. More exactly, we have the following

THEOREM 8.3. *If G is a topological abelian group, then for any complex exponential polynomial $f : G \to \mathbf{C}$ we have $sp(f) = supp\ \hat{f}$.*

PROOF: By Theorem 8.2 it is enough to show, that if the complex exponential polynomial $f : G \to \mathbf{C}$ has the canonical representation

$$f = \sum_{i=1}^{N} m_i p_i,$$

then m_i is in $\tau(f)$ for $i = 1, 2, \ldots, N$. As $\tau(f)$ is a translation invariant linear space, Lemma 4.2 can be applied, and we see, that it is enough to prove our statement if f is of order 1, that is, $f = mp$, where m is a continuous nonzero complex exponential and p is a nonzero complex polynomial. If p is

63

a constant, then the statement is trivial, hence we may suppose, that it is proved for polynomials of degree at most $n - 1$, and p is of degree $n \geq 1$. As in Theorem 5.1, there exists a y in G such that $\Delta_y p$ is not identically zero. As $\Delta_y p$ is a polynomial of degree at most $n - 1$, the equation

$$m(x)\Delta_y p(x) = m(-y)f(x + y) - f(x)$$

shows, that $m\Delta_y p$ is in $\tau(f)$. Hence our statement follows by induction.

The fundamental results about mean periodic functions on topological abelian groups concern with spectral synthesis. We say that *spectral synthesis* holds for a variety in $\mathcal{C}(G)$, if the linear hull of all exponential monomials contained in the variety is dense in this variety. We recall, that an exponential monomial on the topological abelian group G has the form

$$\varphi(x) = a_1(x)^{\alpha_1} a_2(x)^{\alpha_2} \ldots a_k(x)^{\alpha_k} m(x)$$

for all x in G, where $m : G \to \mathbf{C}$ is a continuous exponential, k is a nonnegative integer, $a_1, a_2, \ldots, a_k : G \to \mathbf{R}$ are continuous additive functions and $\alpha_1, \alpha_2, \ldots, \alpha_k$ are nonnegative integers. If $k = 0$ then the above form means $\varphi = m$ and we use the convention $0^0 = 1$. As it has been mentioned in Section 2, we always suppose, that all additive functions in the above representation are taken from a fixed linearly independent set of continuous real additive functions. Then by normal exponential polynomials on G we mean complex linear combinations of functions of the above form.

The first general result on spectral synthesis is the following one, proved in [SCZ1] for all varieties in $\mathcal{C}(\mathbf{R})$.

THEOREM 8.4. *For any variety in $\mathcal{C}(\mathbf{R})$, the linear hull of all exponential monomials contained in the variety, is dense in this variety.*

This result makes it possible to extend the Fourier-transformation of exponential polynomials, defined in Section 7, for all mean periodic functions on the real line.

THEOREM 8.5. *There exists a unique continuous linear operator*

$$M : \mathcal{MP}(\mathbf{R}) \to \mathcal{P}(\mathbf{R})$$

satisfying the properties

(i) $M(\tau_y f) = \tau_y M(f)$

(ii) $M(p) = p$

for all f in $\mathcal{M}\mathcal{P}(\mathbf{R})$, p in $\mathcal{P}(\mathbf{R})$ and y in $\mathbf{R}$.

PROOF: First we prove the uniqueness. By Theorem 8.4 it is enough to show that the properties of M determine M on the set of all exponential polynomials, or on the set of all exponential monomials. Let $m \neq 1$ be any nonzero continuous complex exponential. Then we have

$$M(m) = M[m(-y)\tau_y m] = m(-y)M(\tau_y m) = m(-y)\tau_y M(m),$$

which implies that either $M(m) = 0$ or m is a polynomial. Hence $M(m) = 0$. Suppose that we have proved for $j = 0, 1, \ldots, k-1$ that

$$M[x^j m(x)] = 0$$

for any continuous complex exponential $m \neq 1$. Then we have

$$M[(x+y)^k m(x+y)] = M\left[\sum_{j=0}^{k}\binom{k}{j}x^j y^{k-j}m(x)m(y)\right] =$$

$$= \sum_{j=0}^{k}\binom{k}{j}y^{k-j}m(y)M[x^j m(x)] =$$

$$= m(y)M[x^k m(x)],$$

which implies, as above, that $M[x^k m(x)] = 0$. This proves the uniqueness.

To prove the existence, first we notice, that for any nonzero μ in $\mathcal{M}_c(\mathbf{R})$, the exponential 1 is not contained in the closed linear subspace of $\mathcal{C}(\mathbf{R})$ spanned by all exponential monomials in $V(\mu)$ different from 1. This follows from the results of [SCZ 1, Théoréme 7]. It implies the existence of a measure μ_0 in $\mathcal{M}_c(\mathbf{R})$ such that $\mu_0(1) = 1$ and $\mu_0(\varphi) = 0$ for any exponential monomial $\varphi \neq 1$ in $V(\mu)$.

From this fact it follows, that $x^k m(x) * \mu_0 = 0$ for all positive integers k and exponentials $m \neq 1$ in $V(\mu)$; further $x^k * \mu_0 = x^k$. This shows, that $\varphi * \mu_0$ is a polynomial in $V(\mu)$ for any exponential polynomial φ in $V(\mu)$. On the other hand, it is easy to see that if a polynomial of degree n belongs to $V(\mu)$, then the functions $1, x, x^2, \ldots, x^n$ also belong to $V(\mu)$. (This follows also from Lemma 2.8.) Hence, all polynomials in $V(\mu)$ must have a degree smaller than some fixed positive integer, or else the Stone-Weierstrass theorem would imply, that $V(\mu) = \mathcal{C}(\mathbf{R})$, which is not the case. Now, if f is arbitrary in $V(\mu)$, then by Theorem 8.4 there exist exponential polynomials φ_k in $V(\mu)$ such that $f = \lim \varphi_k$. Then we have $f * \mu_0 = \lim(\varphi_k * \mu_0)$, hence also $f * \mu_0$ is a polynomial.

Suppose now, that f belongs also to some $V(\nu)$ with some nonzero ν. Then $f * \mu_0$ also belongs to $V(\nu)$, and it is a polynomial. Hence we have $f * \mu_0 = f * \mu_0 * \nu_0$. Similarly, $f * \nu_0 = f * \nu_0 * \mu_0$. Hence $f * \mu_0$ does not depend on the special choice of μ. On the other hand, each f in $\mathcal{MP}(\mathbf{R})$ is contained in some $V(\mu)$ with $\mu \neq 0$, and we can define

$$M(f) = f * \mu_0$$

for any nonzero μ in $\mathcal{M}_c(\mathbf{R})$ with $f * \mu = 0$. The continuity and linearity of M follows from the definition; (i) follows from the properties of convolution, and (ii) has been proved.

By the fact, that $f\breve{m}$ is mean periodic for any f in $\mathcal{MP}(\mathbf{R})$ and for any continuous complex exponential m, we may define $\hat{f}$ as follows:

$$\hat{f}(m) = M(f\breve{m})$$

for any nonzero continuous exponential $m : \mathbf{R} \to \mathbf{C}$.

LEMMA 8.6. *The map $f \mapsto \hat{f}$ defined above is linear and has the following properties:*
  (i) $\hat{p}(m) = 0$ *for* $m \neq 1$, $\hat{p}(1) = p$,
  (ii) $(pf)\widehat{} = p\hat{f}(m)$,
  (iii) $(\tau_y f)\widehat{}(m) = m(y)(\tau_y \hat{f})(m)$,
  (iv) $(\breve{f})\widehat{}(m) = [\hat{f}(\breve{m})]\breve{}$,
for any f in $\mathcal{MP}(\mathbf{R})$, for any p in $\mathcal{MP}(\mathbf{R})$ and y in $\mathbf{R}$, whenever pf is mean periodic.

The statements are direct consequences of Lemma 8.5 and the definition.

THEOREM 8.7. *For any f in $\mathcal{MP}(\mathbf{R})$, if $\hat{f} = 0$, then $f = 0$.*

PROOF: From Theorem 8.6 it follows by linearity and continuity, that $\hat{\varphi} = 0$ for all φ in $\tau(f)$. In particular, $\hat{\varphi} = 0$ for any exponential polynomial φ in $\tau(f)$, hence, the inversion theorem 7.3 for exponential polynomials implies that the

66

only exponential polynomial in $\tau(f)$ is 0. Now our statement is a consequence of the spectral synthesis theorem 8.4.

COROLLARY 8.8. *For any f in $\mathcal{MP}(\mathbf{R})$ we have*

$$sp(f) = supp \; \hat{f}.$$

In particular, f is an exponential polynomial if and only if the support of $\hat{f}$ is finite.

COROLLARY 8.9. *For any almost periodic f in $\mathcal{MP}(\mathbf{R})$, the function $\hat{f}$ coincides with the Fourier-transform of f in the sense of the theory of almost periodic functions.*

PROOF: By the results of [KAH2] it follows, that in this case $\tau(f)$ contains only bounded exponential polynomials, that is, only trigonometric polynomials. On the other hand, for trigonometric polynomials our statement is obvious (see [MAA]).

The result of [SCZ1] has been extended in [EHR1] for those varieties on an arbitrary locally compact abelian group, whose annihilator ideal is a principal ideal. Spectral synthesis for any varieties on finitely generated discrete abelian groups has been proved in [LEF].

THEOREM 8.10. *Let G be a finitely generated discrete abelian group. Then the linear hull of all exponential monomials contained in a variety is dense in this variety.*

The extension for any discrete abelian group is due to [ELL2].

THEOREM 8.11. *Let G be a discrete abelian group. Then the linear hull of all exponential monomials contained in a variety is dense in this variety.*

We remark that the result of [LEF] on the spectral synthesis for finitely generated discrete abelian groups is enough for our purposes, that is, we actually don't need the result of [ELL2] for any discrete abelian group. To see this, we need the following theorem.

67

THEOREM 8.12. *Let G be a discrete abelian group, $f : G \to \mathbf{C}$ a function, n a positive integer and α an n-dimensional multi-index. If the restriction of f to any finitely generated subgroup of G is an exponential polynomial of order n and of type α, then f is an exponential polynomial of order n and type α.*

PROOF: For any finitely generated subgroup $F \subset G$ let f_F denote the restriction of f to F. Then we have the canonical representation of f_F in the form:

$$f_F = \sum_{j=1}^{n} p_{j,F} m_{j,F},$$

where $m_{j,F} : F \to \mathbf{C}$ is an exponential, $m_{j,F} \neq m_{i,F}$ for $j \neq i$, and $p_{j,F} : F \to \mathbf{C}$ is a polynomial of degree at most α_j $(i, j = 1, 2, \ldots, n)$. Using the uniqueness of the canonical representations of exponential polynomials, we may suppose that $m_{j,F_1} = m_{j,F_2}$ on F_1 whenever $F_1 \subset F_2$. This means, that there exists the limit of the net $\{m_{j,F}\}$ over the directed set of finitely generated subgroups of G, and it is easy to see, that the limit is an exponential $m_j : G \to \mathbf{C}$ $(j = 1, 2, \ldots, n)$. Now we have

$$f_F = \sum_{j=1}^{n} p_{j,F} m_j,$$

and obviously f is the limit of the net $\{f_F\}$ over the directed set of finitely generated subgroups of G. Hence our statement is a consequence of Theorem 4.6.

Hence if we can prove that the solution of a given system of functional equations restricted to any finitely generated subgroup is an exponential polynomial of fixed order and type, then it is an exponential polynomial of the same order and type on the whole group.

We note, that if G is a discrete abelian group, then $\mathcal{C}(G)$ is the set of all complex valued functions on G equipped with the topology of pointwise convergence, and $\mathcal{M}_c(G)$ is the set of all discrete measures. A function $f : G \to \mathbf{C}$ is mean periodic if and only if there exists a nonzero measure $\mu \in \mathcal{M}_c(G)$ with finite support such that $f * \mu = 0$. In other words, mean periodic functions on discrete abelian groups are solutions of nontrivial difference equations of the form

$$\sum_{i=1}^{n} c_i f(x + y_i) = 0.$$

68

Suppose now, that $\Lambda \subset \mathcal{M}_c(G)$ is given and we consider the system of functional equations

$$f * \mu = 0, \qquad \mu \in \Lambda.$$

We say, that two systems of this form are *equivalent*, if the solution spaces are the same. For any subgroup $F \subset G$ the *restriction* of this system of equations to F is the system

$$f * \mu = 0, \qquad \mu \in \Lambda, \ supp \ \mu \subset F.$$

The following theorem makes it possible to prove equivalence of systems of functional equations using spectral synthesis on finitely generated subgroups only.

THEOREM 8.13. *Let G be a discrete abelian group, $\Lambda_1, \Lambda_2 \subset \mathcal{M}_c(G)$. If the restrictions of the systems of functional equations*

$$f * \mu = 0, \qquad \mu \in \Lambda_i$$

($i = 1, 2$) to any finitely generated subgroup of G are equivalent, then the two systems equivalent.

PROOF: Let V_i denote the respective solution spaces ($i = 1, 2$). Suppose, that there is an $f \in V_1$ with $f \notin V_2$. This means, that we have

$$f * \mu = 0, \qquad \mu \in \Lambda_1$$

and there exists $\nu \in \Lambda_2$ with $f * \nu \neq 0$. This latter relation means, that there exists an $x_0 \in G$ with $f * \nu(x_0) \neq 0$. Consider the subgroup F of G generated by x_0 and *supp* ν. This is a finitely generated subgroup of G and the restriction of f to F is obviously a solution of the restriction of the system

$$f * \mu = 0, \qquad \mu \in \Lambda_1$$

to F. On the other hand, the restriction of f to F is not a solution of

$$f * \nu = 0$$

on F, because $x_0 \in F$. Hence, the restriction of f to F is not a solution of the restriction of

$$f * \mu = 0, \qquad \mu \in \Lambda_2$$

to F, which is a contradiction.

REFERENCES 8.14.

For further results and references concerning mean periodic functions and spectral synthesis see [BEN], [EHR1], [EHR2], [ELL1], [ELL2], [GIL1], [GIL2], [KAH1], [KAH2], [LAI1], [LEF], [SCZ], [SZÉ18].

CHAPTER 3

APPLICATIONS FOR FUNCTIONAL EQUATIONS

9. Functional equations for polynomials

In this section we deal with functional equations, which serve to characterize polynomials, and we give some applications. Most of the results of this section are classical and well-known; although in some cases we give here a new proof based on the previous results. The literature concerning the present results is enormous; see the reference list at the end of this section.

The following well-known theorem shows that the Fréchet equation:

$$(9.1) \qquad \Delta_{y_1,y_2,\ldots,y_{n+1}} f(x) = 0$$

characterizes polynomials of degree at most n.

THEOREM 9.1. *Let G be a commutative semigroup with identity, S a commutative group and n a nonnegative integer. Let the multiplication by $n!$ be bijective in S. The function $f : G \to S$ is a solution of the Fréchet equation (9.1) if and only if f is a polynomial of degree at most n.*

PROOF: The sufficiency follows from Lemma 1.4. To prove the necessity, we suppose that f satisfies (9.1). Then $\Delta_{y_1,y_2,\ldots,y_n} f$ is constant, and we define

$$A_n(y_1, y_2, \ldots, y_n) = \frac{1}{n!} \Delta_{y_1,y_2,\ldots,y_n} f(x)$$

for any $y_1, y_2, \ldots, y_n$ in G. Obviously A_n is symmetric, and we have by Lemma 1.4

$$\Delta_{y_1,y_2,\ldots,y_n}(f - A_n)(x) = n! A_n(y_1, y_2, \ldots, y_n) - n! A_n(y_1, y_2, \ldots, y_n) = 0$$

for all $x, y_1, y_2, \ldots, y_n$ in G, hence our statement follows by induction.

We note that we need an identity in G only to derive, that the vanishing of $\Delta_{y_1,y_2,\ldots,y_n,y_{n+1}} f$ implies, that $\Delta_{y_1,y_2,\ldots,y_n} f$ is constant. If we use this latter condition instead of (9.1), then Theorem 9.1 remains valid for an arbitrary commutative semigroup G.

The following lemma allows us to restrict ourselves in most cases to linear space valued functions.

LEMMA 9.2. *Let G be a commutative semigroup with identity, S, T commutative groups, n a nonnegative integer and suppose, that multiplication by $n!$ is bijective in S, and $Hom(S, T)$ is a separating family for S. Then a function $f : G \to S$ is a polynomial of degree at most n if and only if $\varphi \circ f$ is a polynomial of degree at most n for any additive function $\varphi : S \to T$.*

PROOF: If $f : G \to S$ is a polynomial of degree at most n with the canonical representation

$$f = \sum_{k=0}^{n} A_k^*$$

and $\varphi : S \to T$ is an additive function, then obviously we have

$$\varphi \circ f = \sum_{k=0}^{n} \varphi \circ A_k^* = \sum_{k=0}^{n} (\varphi \circ A_k)^*,$$

which shows that $\varphi \circ f$ is a polynomial of degree at most n. On the other hand, if $\varphi \circ f$ is a polynomial of degree at most n for any additive function $\varphi : S \to T$, then we have for all $x, y_1, y_2, \ldots, y_{n+1}$ in G

$$\varphi\left(\Delta_{y_1, y_2, \ldots, y_{n+1}} f(x)\right) = \Delta_{y_1, y_2, \ldots, y_{n+1}} (\varphi \circ f)(x) = 0$$

and hence, by the separating property of $Hom(S, T)$ it follows

$$\Delta_{y_1, y_2, \ldots, y_{n+1}} f(x) = 0,$$

which was to be proved, by Theorem 9.1.

Theorem 9.1 shows, that in dealing with polynomials of arbitrary degree, we should suppose, that their range is contained in an abelian group, in which multiplication by any positive number is bijective, that is, their range is contained in a divisible and torsion-free abelian group. As these groups are linear spaces over the rationals, from now on we may restrict ourselves to polynomials with values in linear spaces over any field of characteristic zero. Lemma

9.2 shows that on abelian groups further reduction is possible. Namely, if a functional equation for functions from G into S is "linear" in the sense, that $\varphi \circ f$ is a solution of it for any complex additive function φ on S, whenever $f : G \to S$ is a solution, and all complex solutions on G are polynomials of degree at most n, then all solutions from G into S are polynomials of degree at most n, supposing, that $Hom(S, \mathbf{C})$ is a separating family. But this is obviously the case, if S is a linear space over the rationals. On the other hand, for complex valued functions on abelian groups we can often use the results of Section 8 on spectral synthesis, as it can be seen by the following theorems.

THEOREM 9.3. *Let G be an abelian group and S a linear space over the rationals. The function $f : G \to S$ is a polynomial if and only if it satisfies*

$$(9.2) \qquad \qquad \Delta_y f(x) = 0$$

for some nonnegative integer n, and for all x, y in G. In this case the degree of f is at most n.

PROOF: We have to prove the sufficiency only. As $Hom(S, \mathbf{C})$ is a separating family for S and $\varphi \circ f$ is a solution of (9.2) together with f for any complex additive function φ on S, we may suppose by Lemma 9.2, that $S = \mathbf{C}$. By Theorem 8.12 we may suppose that G is finitely generated and by spectral synthesis any solution f of (9.2) is the pointwise limit of (normal) exponential polynomial solutions of (9.2). First we determine the spectrum of $\tau(f)$. The nonzero complex exponential m belongs to the spectrum if and only if

$$(m(y) - 1)^{n+1} = \sum_{k=0}^{n+1} \binom{n+1}{k} (-1)^{n+1-k} m(y)^k = 0$$

holds for any y in G, hence $m = 1$. Thus any exponential polynomial solution of (9.2) is a polynomial, and, by Lemma 1.4, it is of degree at most n. Then our statement is a consequence of Theorem 2.4.

COROLLARY 9.4. *Let G be an abelian group and S a linear space over the rationals. The function $f : G \to S$ is a homogeneous polynomial if and only if it satisfies*

$$(9.3) \qquad \qquad \Delta_y f(x) = n! f(y)$$

for some nonnegative integer n, and for all x, y in G. In this case the degree of f is n, whenever it is not identically zero.

PROOF: Theorem 9.3 implies that f is a polynomial of degree at most n. Then the statement is a consequence of Theorem 2.5.

The same method can be applied for the more general functional equation

$$(9.4) \qquad f(x) + \sum_{i=1}^{n} f_i(\varphi_i(x) + \psi_i(y)) = 0,$$

where φ_i, ψ_i are homomorphisms of the underlying group, which is the common domain of the unknown functions f, f_i. Equation (9.4) is a common generalization of classical functional equations, as e.g.

the Pexider equation:

$$(9.5) \qquad f(x + y) = g(x) + h(y);$$

the Jensen equation:

$$(9.6) \qquad 2f\left(\frac{x+y}{2}\right) = f(x) + f(y);$$

and the square-norm equation:

$$(9.7) \qquad f(x + y) + f(x - y) = 2f(x) + 2f(y).$$

Obviously, a "pexiderization" - that is, putting different functions for different appearances of f - is possible also in the latter two cases.

In general, (9.4) can be reduced to (9.1).

THEOREM 9.5. *Let G be a commutative semigroup, S a commutative group, n a nonnegative integer, φ_i, ψ_i additive functions from G into G and let $rg(\varphi_i) \subseteq rg(\psi_i)$ $(i = 1, 2, \ldots, n+1)$. If the functions $f, f_i : G \to S$ $(i = 1, 2, \ldots, n+1)$ satisfy (9.4), then f satisfies (9.1).*

PROOF: We call a function $f : G \to S$ a function of degree n, if there exist functions $f_i : G \to S$ and additive functions φ_i, ψ_i from G into G with $rg(\varphi_i) \subseteq rg(\psi_i)$ $(i = 1, 2, \ldots, n+1)$ such that (9.4) holds for all x, y in G. It is obvious, that f is of degree 0 if and only if it is constant. Now we prove, that if f is of degree n $(n \geq 1)$, then $\Delta_t f$ is of degree $n-1$, for any t in G. Let t be arbitrary and we choose an s in G with the property that

$$\varphi_{n+1}(t) + \psi_{n+1}(s) = 0.$$

The existence of an s with this property follows from the assumption on the ranges of the given homomorphisms. If we put $x + t$ for x and $y + s$ for y in (9.4), then we have for all x, y in G

$$\Delta_t f(x) + \sum_{i=1}^{n} \Delta_{\varphi_i(t)+\psi_i(s)} f_i(\varphi_i(x) + \psi_i(y)) = 0,$$

which proves our statement. Hence $\Delta_{y_1,y_2,\dots,y_{n+1}} f$ vanishes for all $y_1, y_2, \dots, y_{n+1}$ in G, and our theorem is proved.

By the previous results and using Theorem 2.5 we can describe the well-known general solutions of equations (9.5) - (9.7).

THEOREM 9.6. *Let G be a commutative semigroup with identity, and let S be a commutative group. The functions $f, g, h : G \to S$ are solutions of (9.5) if and only if there exists an additive function $a : G \to S$ and there exist constants b, c in S with*

$$f(x) = a(x) + b + c, \qquad g(x) = a(x) + b, \qquad h(x) = a(x) + c \qquad (x \in G).$$

THEOREM 9.7. *Let G be a commutative semigroup with identity in which multiplication by 2 is bijective and let S be a commutative group. The function $f : G \to S$ is a solution of (9.6) if and only if there exists an additive function $a : G \to S$ and a constant b in S with*

$$f(x) = a(x) + b \qquad (x \in G).$$

THEOREM 9.8. *Let G, S be a commutative groups and suppose that multiplication by 2 is bijective in S. The function $f : G \to S$ is a solution of (9.7) if and only if there exists a symmetric biadditive function $B : G \times G \to S$ with*

$$f(x) = B^*(x) \qquad (x \in G).$$

Using the results of Section 3 several regularity-type consequences of the above theorems can be formulated for equations of the form (9.1) - (9.7). We give here a possible local version.

THEOREM 9.9. *Let G be a topological abelian group generated by any neighborhood of zero and let X be a locally convex topological vector space. Let n be a nonnegative integer, φ_i, ψ_i additive functions from G into G with $rg(\varphi_i) \subseteq rg(\psi_i)$ and let $f, f_i : G \to X$ be functions, satisfying (9.4) $(i = 1, 2, \dots, n + 1)$. Then any of the following conditions implies that f is a polynomial of degree at most n:*
  (i) f is continuous at a point ;
  (ii) f is bounded on a nonvoid open set ;

74

(iii) *G is locally compact and f is bounded on a measurable set of positive measure ;*

(iv) *G is locally compact, X is locally bounded and f is measurable on a measurable set of positive measure.*

As a further application we consider the functional equation

$$(9.8) \qquad f(x)g(y) = \prod_{i=1}^{n} h(a_i x + b_i y).$$

Equation (9.8) plays a role in the characterization theory of probability distributions and has been dealt with e.g. in [ACZ7], [BAK3], [BAK4], [COR1], [COR2], [COR3], [ECS1], [ECS2], [GHU], [HLL], [KAG], [KAN], [KHA1], [KHA2], [LAJ1], [LAJ2], [LAJ3], [LAJ4], [LUK], [RAO], [SZÉ1], [SZÉ7]. In order to apply our previous results, we need the following lemma (see [BAK3]).

LEMMA 9.10. *Let n be a positive integer, a_i, b_i nonzero real numbers with $a_i b_j \neq a_j b_i$ for $i \neq j$ $(i, j = 1, 2, \ldots, n)$, $f, g, h_i : \mathbf{R} \to \mathbf{R}$ functions satisfying (9.8) for all x, y in $\mathbf{R}$, and suppose, that there exists a measurable set $D \subseteq \mathbf{R}^2$ of positive measure such that $f(x)g(y) \neq 0$ for all (x, y) in D. Then f, g, h_i are never vanishing.*

PROOF: By assumption, for at least one y_0 in $\mathbf{R}$, the set $A = \{x \in \mathbf{R} : (x, y_0) \in D\}$ is of positive measure and $f(x) \neq 0$ for $x \in A$. Similarly, there exists a measurable subset B of positive measure, with $g(y) \neq 0$ for $y \in B$. If $\tilde{A}, \tilde{B}$ denote the set of points of unit density of A and B, resp., then we have

$$f(x)g(y) \neq 0$$

for all x in $\tilde{A}$ and y in $\tilde{B}$. As $a_i \neq 0$, the set of density points of $a_i A$ is $a_i \tilde{A}$ for $i = 1, 2, \ldots, n$. Similarly, the set of density points of $b_i B$ is $b_i \tilde{B}$. By a result of [HAL] we conclude that $a_i \tilde{A} + b_i \tilde{B}$ is a nonvoid open subset of $\mathbf{R}$ for $i = 1, 2, \ldots, n$ and $h_i(u) \neq 0$ whenever $u \in a_i \tilde{A} + b_i \tilde{B}$. Now let

$$U = \{(x, y) \in \mathbf{R}^2 : a_i x + b_i y \in a_i \tilde{A} + b_i \tilde{B} \text{ for } i = 1, 2, \ldots, n\}.$$

Then U is open, and $U \neq \emptyset$ since $U \supset \tilde{A} \times \tilde{B}$. Also $f(x)g(y) \neq 0$ for (x, y) in U. Hence there exist intervals I, J in $\mathbf{R}$ both of length $l > 0$ such that

$$f(x)g(y) \neq 0$$

for all x in I and y in J. It follows from (9.8) that $h_i(u) \neq 0$ whenever $u \in a_i I + b_i J$ (an interval) for $i = 1, 2, \ldots, n$. Now for each $i = 1, 2, \ldots, n$ let

$$S_i = \{(x, y) \in \mathbf{R}^2 : a_i x + b_i y \in a_i I + b_i J\}$$

and let $S = \cap_{i=1}^n S_i$. Then we have

$$f(x)g(y) \neq 0$$

whenever (x, y) is in S. Notice, that each S_i is a "strip" parallel to the line $\{(x, y) : a_i x + b_i y = 0\}$ and is the smallest such strip containing $I \times J$. Moreover none of these strips is parallel to either co-ordinate axis. Hence, if

$$I' = \{x : (x, y) \in S \ for \ some \ y \in \mathbf{R}\}$$

and

$$J' = \{y : (x, y) \in S \ for \ some \ x \in \mathbf{R}\},$$

then I' and J' are intervals with the same center as I and J respectively, but with lengths greater than l. Moreover, $f(x)g(y) \neq 0$ when $x \in I'$ and $y \in J'$. In fact, we have shown the existence of a constant $\omega > 1$ depending only on $a_1, b_1, a_2, b_2, \ldots, a_n, b_n$ such that if $f(x)g(y) \neq 0$ for all (x, y) belonging to some square of side l, then $f(x)g(y) \neq 0$ for all (x, y) belonging to the square with the same center, but with side ωl. It follows by induction that $f(x)g(y) \neq 0$ for all (x, y) in $\mathbf{R}^2$ and hence $f(x)g(y)h_i(u) \neq 0$ for all x, y, u in $\mathbf{R}$, which completes the proof.

THEOREM 9.11. *Let n be a positive integer, $f, g, h_i : \mathbf{R} \to \mathbf{R}$ functions, which are not almost everywhere zero and satisfy (9.8) for all x, y in $\mathbf{R}$, and let a_i, b_i be real numbers with $a_i b_j \neq a_j b_i$ for $i \neq j$ $(i, j = 1, 2, \ldots, n)$. Then there exist $A_k, B_k, C_{i,k} : \mathbf{R}^k \to \mathbf{R}$ k-additive and symmetric functions, and α, β, γ_i nonzero real numbers $(i, k = 1, 2, \ldots, n)$ such that*

$$f = \alpha \exp \sum_{k=1}^n A_k^*, \qquad g = \beta \exp \sum_{k=1}^n B_k^*, \qquad h_i = \gamma_i \exp \sum_{k=1}^n C_{i,k}^*,$$

holds for $i = 1, 2, \ldots, n$.

PROOF: By the previous lemma, f, g, h_i are never vanishing. We show, that $f/f(0)$ is positive. For $n = 1$ this is trivial. Let $n \geq 2$, let $t \in \mathbf{R}$ be arbitrary and we choose $s \in \mathbf{R}$ with $a_n t + b_n s = 0$. Substituting $x + t$ for x and $y + s$ for y in (9.8), and dividing the new equation by (9.8) we obtain

$$\frac{f(x+t)}{f(x)} \frac{g(y+s)}{g(y)} = \prod_{i=1}^{n-1} \frac{h_i(a_i x + b_i y + a_i t + b_i s)}{h_i(a_i x + b_i y)}$$

for all x, y, t, s in $\mathbf{R}$. By induction it follows

$$\frac{f(x+t)f(0)}{f(x)f(t)} > 0$$

for all x, t in $\mathbf{R}$, which yields our statement by substituting $x/2$ for x and t. Similarly, we can show, that $g/g(0)$ and $h_i/h_i(0)$ are positive for $i = 1, 2, \ldots, n$. Hence (9.8) implies

$$\frac{f(x)}{f(0)} \frac{g(y)}{g(0)} = \prod_{i=1}^{n} \frac{h_i(a_i x + b_i y)}{h_i(0)}$$

for all x, y in $\mathbf{R}$, that is, the functions $f/f(0), g/g(0)$ and $h_i/h_i(0)$ ($i = 1, 2, \ldots, n$) satisfy an equation of the form (9.4) under the conditions of Theorem 9.1 and Theorem 9.5, where $G = \mathbf{R}$ and S is the multiplicative group of positive real numbers. Then, by Theorem 9.1 our theorem is proved.

REFERENCES 9.12.

For references see [ACZ4], [ACZ6], [ACZ7], [ACZ8], [ACZ13], [ACZ14], [ACZ15], [ACZ16], [ACZ18], [ALE], [ANG1], [ANG2], [BAK3], [BAK4], [CAU], [COR1], [COR2], [COR3], [DAR], [DJO1], [DJO3], [ECS1], [ECS2], [FRE1], [FRE2], [FRE3], [FRE4], [GHE1], [GHE3], [GHU], [GIR1], [HAM12], [HAS4], [HLL], [HOS1], [HOS2], [ING], [KAG], [KAN], [KCZ4], [KEM1], [KEM3], [KEN], [KES], [KHA1], [KHA2], [KUR3], [LAJ1], [LAJ2], [LAJ3], [LAJ4], [LIJ1], [LIJ2], [LIJ3], [LUK], [MAK], [MAZ], [MCK1], [MCK3], [OST], [PEX], [POP1], [POP2], [RAO], [RES], [SWI1], [SWI2], [SZÉ1], [SZÉ2], [SZÉ3], [SZÉ7], [SZÉ8], [SZÉ12], [SZÉ13].

10. The Levi-Civitá functional equation

In this section we deal with finite dimensional translation invariant linear function spaces on topological abelian groups. This problem concerns the finite dimensional representations of topological abelian groups. We note, that some of the results of this section can be (and has been) generalized for commutative semigroups and even for some groupoids (see e.g. [MCK7]). However, here we deal with the case of groups only, in order to apply the tools given by spectral synthesis.

THEOREM 10.1. *Any finite dimensional translation invariant linear space of continuous complex valued functions on a topological abelian group consists of normal exponential polynomials.*

PROOF: Obviously we may suppose that the group G is discrete and finitely generated (see Theorem 8.12). By spectral synthesis, any variety is the closed linear hull of all exponential monomials belonging to this variety. As any finite dimensional subspace is closed, our theorem follows.

By using the results on the regularity properties of exponential polynomials in Section 5 - especially Theorem 5.10 - we get the following theorem.

THEOREM 10.2. *Any finite dimensional translation invariant linear space of measurable complex valued functions on a locally compact abelian group consists of normal exponential polynomials.*

Theorem 10.1 implies that if G is an abelian group, then any function $f : G \to \mathbf{C}$, satisfying the functional equation

$$(10.1) \qquad f(x + y) = \sum_{i=1}^{n} g_i(x) h_i(y)$$

for some positive integer n and functions $g_i, h_i : G \to \mathbf{C}$ $(i = 1, 2, \ldots, n)$, is a normal exponential polynomial of order at most n. It is easy to see also the converse, that is, any normal exponential polynomial f satisfies an equation of the form (10.1). Functional equation (10.1) is called Levi-Civitá equation and has been studied by several authors under various assumptions. The first appearances of (10.1) can be found in [LEV], [MAG], [STA], [STE], where (10.1) has been solved under differentiability conditions. This method can be found in [ACZ7]. Later the differentiability condition has been weakened in [KEM1], [KEM4], [SAT]. A general method for solving (10.1) is given in [VIN3] (see also [SAK]), where special cases of (10.1) has been solved. In [MCK6], [MCK7] (see also [LES1], [LES2]) it has been proved that all solutions f of (10.1) are normal exponential polynomials, by reducing it to a matrix equation (see also [KCZ1], [KCZ2], [KCZ3], [KCZ4], [SCW], [SZÉ6], [SZÉ7] [SZÉ9]). The complete description of all (nondegenerate) solutions of (10.1) for any n can be found in [SZÉ24].

It is obvious, that if the functions $h_1, h_2, \ldots, h_n$ are linearly independent, then $g_1, g_2, \ldots, g_n$ are linear combinations of translates of f, hence they are normal exponential polynomials of order at most n too, moreover, they are built up from the same additive and exponential functions, as f. If either the functions $g_1, g_2, \ldots, g_n$ or the functions $h_1, h_2, \ldots, h_n$ are linearly dependent, then the number n in (10.1) can be reduced and in this case the general solution of (10.1) contains arbitrary functions. We shall call this case of (10.1) *degenerate*. Thus, in the nondegenerate case all the functions $f, g_1, g_2, \ldots, g_n$ and $h_1, h_2, \ldots, h_n$ are normal exponential polynomials of order at most n, built up from the same additive and exponential functions; that is, they have the form

$$f(x) = \sum_{j=1}^{k} P_j(a_{j,1}(x), a_{j,2}(x), \ldots, a_{j,n_j-1}(x))m_j(x),$$

(10.2)
$$g_i(x) = \sum_{j=1}^{k} Q_{i,j}(a_{j,1}(x), a_{j,2}(x), \ldots, a_{j,n_j-1}(x))m_j(x),$$

$$h_i(x) = \sum_{j=1}^{k} R_{i,j}(a_{j,1}(x), a_{j,2}(x), \ldots, a_{j,n_j-1}(x))m_j(x),$$

where $k, n_1, n_2, \ldots, n_k$ are positive integers, $m_1, m_2, \ldots, m_k$ are different nonzero complex exponentials (the spectrum of (10.1)), further $\{a_{j,1}, a_{j,2} \ldots, a_{j,n_j-1}\}$ are sets of linearly independent real additive functions for $j = 1, 2, \ldots, k$, and $P_j, Q_{i,j}, R_{i,j} : \mathbf{C}^{n_j-1} \to \mathbf{C}$ are complex polynomials of degree at most $n_j - 1$ in $n_j - 1$ variables for $i = 1, 2, \ldots, n$ and $j = 1, 2, \ldots, k$. By Lemma 4.5

we have $n_1 + n_2 + \cdots + n_k = n$. Here the number n_j is called the *multiplicity* of the exponential m_j. Of course, the polynomials $P_j, Q_{i,j}, R_{i,j}$ in (10.2) must satisfy a lot of extra relations in order that (10.2) forms a - nondegenerate - solution of (10.1). To determine these relations we make use of the following theorem.

THEOREM 10.3. *Let G be an abelian group, n a positive integer and $f, g_i, h_i :$ $G \to \mathbf{C}$ normal exponential polynomials, where $\{g_1, g_2, \ldots, g_n\}$ and also $\{h_1, h_2, \ldots, h_n\}$ are linearly independent $(i = 1, 2, \ldots n)$. The functions f, g_i, h_i form a nondegenerate solution of (10.1) if and only if for any nonzero complex exponentials $\mu, \nu : G \to \mathbf{C}$ and for any elements x, y in G we have*

$$(10.3) \qquad \sum_{i=1}^{n} \hat{g}_i(\mu)(x)\hat{h}_i(\nu)(y) = \begin{cases} \hat{f}(\mu)(x+y) & for \ \mu = \nu \\ 0 & for \ \mu \neq \nu. \end{cases}$$

PROOF: Taking the Fourier-transform of both sides of (10.1) as functions of x we get

$$\mu(y)\hat{f}(\mu)(x+y) = \sum_{i=1}^{n} \hat{g}_i(\mu)(x)h_i(y)$$

for any nonzero complex exponential μ and for any x, y in G. Now taking the Fourier-transform of both sides as functions of y we have

$$\hat{\mu}(\nu)\hat{f}(\mu)(x+y) = \sum_{i=1}^{n} \hat{g}_i(\mu)(x)\hat{h}_i(\nu)(y)$$

for any nonzero complex exponentials μ, ν and for any x, y in G, which is (10.3). To prove the converse, we use Theorem 7.3 and write

$$\sum_{i=1}^{n} g_i(x)h_i(y) = \sum_{i=1}^{n}\sum_{\mu}\sum_{\nu} \hat{g}_i(\mu)(x)\mu(x)\hat{h}_i(\nu)(y)\nu(y) =$$

$$= \sum_{\mu}\sum_{\nu}\left[\sum_{i=1}^{n} \hat{g}_i(\mu)(x)\hat{h}_i(\nu)(y)\right]\mu(x)\nu(y) = \sum_{\mu} \hat{f}(\mu)(x+y)\mu(x) = f(x+y),$$

which proves our theorem.

Now, using Theorem 10.3 we give a necessary and sufficient condition in terms of the coefficients of the complex polynomials $P_i, Q_{i,j}, R_{i,j}$ in (10.2) in order that f, g_i, h_i is a nondegenerate solution of (10.1). This characterization theorem leads to a very simple algorithm for the solution of (10.1) for arbitrary n, which is similar to the well-known algorithm for the solution of higher order linear differential equations with constant coefficients. We illustrate this method on some classical equations.

We introduce some notation. Let $k, n, n_1, n_2, \ldots, n_k$ be positive integers with $n_1 + n_2 + \cdots + n_k = n$ and let for $j = 1, 2, \ldots, k$ the complex polynomials $P_j, Q_{i,j}$ of $n_j - 1$ variables and of degree at most $n_j - 1$ be given $(i = 1, 2, \ldots, n; j = 1, 2, \ldots, k)$. For any $j = 1, 2, \ldots, k$ and for arbitrary multi-indices $I_j = (i_1, i_2, \ldots, i_{n_j-1})$, $J_j = (j_1, j_2, \ldots, j_{n_j-1})$ we define the $n_j \times n_j$ matrix $M_j(P; I_j, J_j)$ and the $n_j \times n$ matrix $N_j(Q; I_j)$ as follows: for any choice of $p, q = 0, 1, \ldots, n_j - 1$ the $(n_j - p, n_j - q)$-th element of $M_j(P; I_j, J_j)$ is given by

$$M_j(P; I_j, J_j)_{n_j-p, n_j-q} =$$

$$= \begin{cases} \frac{1}{p! q!} \partial_{i_1} \ldots \partial_{i_p} \partial_{j_1} \ldots \partial_{j_q} P_j(0, \ldots, 0) & \text{for } p + q < n_j \\ 0 & \text{for } p + q \geq n_j, \end{cases}$$

and for any choice of $p = 1, 2, \ldots, n_j$, $q = 1, 2, \ldots, n$ the (p, q)-th element of $N_j(Q; I_j)$ is given by

$$N_j(Q; I_j)_{p,q} = \frac{1}{(n_j - p)!} \partial_{i_1} \ldots \partial_{i_{n_j-p}} Q_{q,p}(0, \ldots, 0).$$

Then we define the following two $n \times n$ matrices $M(P; I_1, \ldots, I_k, J_1, \ldots, J_k)$, $N(Q; I_1, \ldots, I_k)$ as the block matrices

$$M(P; I_1, \ldots, I_k, J_1, \ldots, J_k) =$$

$$= \begin{pmatrix} M_1(P; I_1, J_1) & \cdots & \cdots & \cdots \\ \cdots & M_2(P; I_2, J_2) & \cdots & \cdots \\ \vdots & \vdots & \ddots & \vdots \\ \cdots & \cdots & \cdots & M_k(P; I_k, J_k) \end{pmatrix}$$

(the matrices $M_j(P; I_j, J_j)$ are along the main diagonal, all other elements being zero); and

$$N(Q; I_1, \ldots, I_k) = \begin{pmatrix} N_1(Q; I_1) \\ N_2(Q; I_2) \\ \vdots \\ N_k(Q; I_k) \end{pmatrix}.$$

Our main theorem follows.

THEOREM 10.4. *Let G be an abelian group, n a positive integer and $f, g_i, h_i :$ $G \to \mathbf{C}$ functions $(i = 1, 2, \ldots, n)$, where both the sets $\{g_1, g_2, \ldots, g_n\}$ and $\{h_1, h_2, \ldots, h_n\}$ are linearly independent. The functions f, g_i, h_i $(i = 1, 2, \ldots, n)$ form a nondegenerate solution of (10.1) if and only if there exist positive integers $k, n_1, n_2, \ldots, n_k$ with $n_1 + n_2 + \cdots + n_k = n$; there exist different nonzero complex exponentials $m_1, m_2, \ldots, m_k$; there exist linearly independent sets $\{a_{j,1}, a_{j,2}, \ldots, a_{j,n_j-1}\}$ of complex additive functions $(j = 1, 2, \ldots, k)$ and there exist complex polynomials $P_j, Q_{i,j}, R_{i,j} : \mathbf{C}^{n_j-1} \to \mathbf{C}$ $(i = 1, 2, \ldots n; j = 1, 2, \ldots, k)$ in $n_j - 1$ complex variables and of degree at most $n_j - 1$ such that (10.2) holds for all x, y in G, further*

$$(10.4) \qquad M(P; I_1, \ldots, I_k, J_1, \ldots, J_k) =$$

$$= N(Q; I_1, \ldots, I_k)N(R; J_1, \ldots, J_k)^T$$

holds for any choice of the multi-indices I_j, J_j in $\mathbf{N}^{n_j-1}$ $(j = 1, 2, \ldots, k)$. (Here T denotes the transpose of a matrix).

PROOF: The statement is a direct consequence of the previous theorems. Namely, equation (10.3) is equivalent to the system (10.4), which can easily be checked by substituting the polynomials $P_j, Q_{i,j}, R_{i,j}$ in (10.2) into (10.3).

Actually, (10.4) is equivalent to the system of equations

$$(10.4a) \qquad N_p(Q; I_p)N_q(R; J_q)^T = \begin{cases} 0 & \text{for } p \neq q, \\ M_p(P; I_p, J_p) & \text{for } p = q, \end{cases}$$

for any $p, q = 1, 2, \ldots, k$ and for any choice of multi-indices I_p in $\mathbf{N}^{n_p-1}$, J_q in $\mathbf{N}^{n_q-1}$. This is also a system of equations for the coefficients of the polynomials $P_j, Q_{i,j}, R_{i,j}$. Hence to determine all nondegenerate solutions of (10.1) for a given value of n is equivalent to determine for all given decompositions of n into a sum of positive integers $n_1, n_2, \ldots, n_k$ those solutions of the system of scalar equations (10.4) or (10.4a), for which both the corresponding sets $\{g_1, g_2, \ldots, g_n\}$ and $\{h_1, h_2, \ldots, h_n\}$ in (10.2) are linearly independent.

APPLICATIONS 10.5.

Now we illustrate our method on the solution of some special cases of (10.1). Some of these cases have been treated by several methods. Here we show that our approach provides a unified treatment for all cases.

To avoid notational complications we restrict ourselves to the case $n \leq 4$ which is quite general to include well-known special cases, but the most general solution has not been determined yet. First, we consider the functional equation

$$(10.5) \qquad f(x + y) = g_1(x)h_1(y) + g_2(x)h_2(y) + g_3(x)h_3(y) + g_4(x)h_4(y),$$

and we suppose that the complex valued functions $f, g_1, g_2, g_3, g_4, h_1, h_2, h_3, h_4$ form a nondegenerate solution of it. In particular, we consider the special cases

$$h_1 = g_2 = 1, \qquad f = g_1 = h_2, \qquad g_3 = h_4, \qquad g_4 = h_3,$$

which corresponds to the equation

$$(10.5a) \qquad f(x + y) = f(x) + f(y) + g(x)h(y) + g(y)h(x),$$

and

$$h_1 = g_2 = 1, \qquad f = g_1 = h_2, \qquad g_3 = h_3, \qquad g_4 = h_4,$$

which corresponds to the equation

$$(10.5b) \qquad f(x + y) = f(x) + f(y) + g(x)g(y) + h(x)h(y).$$

In the cases $(10.5a)$ and $(10.5b)$ the spectrum contains the exponential $m = 1$. In general there are five possibilities for the spectrum of (10.5) corresponding to the following decompositions of the number 4

(i) $4 = 4$
(ii) $4 = 3 + 1$
(iii) $4 = 2 + 2$
(iv) $4 = 2 + 1 + 1$
(v) $4 = 1 + 1 + 1 + 1.$

In the case (i) $sp(V) = \{m\}$ with $mult(m) = 4$ hence the general nondegenerate solution is

$$f(x) = \sum_{p,q,r=1}^{3} \alpha_{p,q,r} a_p(x) a_q(x) a_r(x) m(x) + \sum_{p,q=1}^{3} \alpha_{p,q} a_p(x) a_q(x) m(x) +$$

$$+ \sum_{p=1}^{3} \alpha_p a_p(x) m(x) + \alpha m(x),$$

83

$$g_i(x) = \sum_{p,q,r=1}^{3} \beta_{p,q,r}^{(i)} a_p(x) a_q(x) a_r(x) m(x) + \sum_{p,q=1}^{3} \beta_{p,q}^{(i)} a_p(x) a_q(x) m(x) +$$

$$+ \sum_{p=1}^{3} \beta_p^{(i)} a_p(x) m(x) + \beta^{(i)} m(x),$$

$$h_i(x) = \sum_{p,q,r=1}^{3} \gamma_{p,q,r}^{(i)} a_p(x) a_q(x) a_r(x) m(x) + \sum_{p,q=1}^{3} \gamma_{p,q}^{(i)} a_p(x) a_q(x) m(x) +$$

$$+ \sum_{p=1}^{3} \gamma_p^{(i)} a_p(x) m(x) + \gamma^{(i)} m(x),$$

$(i = 1, 2, 3, 4)$ where m is a nonzero complex exponential, a_1, a_2, a_3 are linearly independent complex additive functions and all constants are symmetric in the subscripts and are subjected to the conditions

$$\begin{pmatrix} \beta_{p,q,r}^{(1)} & \beta_{p,q,r}^{(2)} & \beta_{p,q,r}^{(3)} & \beta_{p,q,r}^{(4)} \\ \beta_{p,q}^{(1)} & \beta_{p,q}^{(2)} & \beta_{p,q}^{(3)} & \beta_{p,q}^{(4)} \\ \beta_p^{(1)} & \beta_p^{(2)} & \beta_p^{(3)} & \beta_p^{(4)} \\ \beta^{(1)} & \beta^{(2)} & \beta^{(3)} & \beta^{(4)} \end{pmatrix} \begin{pmatrix} \gamma_{s,t,u}^{(1)} & \gamma_{s,t}^{(1)} & \gamma_s^{(1)} & \gamma^{(1)} \\ \gamma_{s,t,u}^{(2)} & \gamma_{s,t}^{(2)} & \gamma_s^{(2)} & \gamma^{(2)} \\ \gamma_{s,t,u}^{(3)} & \gamma_{s,t}^{(3)} & \gamma_s^{(3)} & \gamma^{(3)} \\ \gamma_{s,t,u}^{(4)} & \gamma_{s,t}^{(4)} & \gamma_s^{(4)} & \gamma^{(4)} \end{pmatrix} =$$

$$= \begin{pmatrix} 0 & 0 & 0 & \alpha_{p,q,r} \\ 0 & 0 & 3\alpha_{s,p,q} & \alpha_{p,q} \\ 0 & 3\alpha_{s,p,t} & 2\alpha_{s,p} & \alpha_p \\ \alpha_{s,t,u} & \alpha_{s,t} & \alpha_s & \alpha \end{pmatrix}$$

for all $p, q, r, s, t, u = 1, 2, 3$.

In the case $(10.5a)$ we have

$$f(x) = \frac{1}{3} \sum_{p,q,r=1}^{3} (\beta_{p,q}\gamma_r + \gamma_{p,q}\beta_r) a_p(x) a_q(x) a_r(x) +$$

$$+ \frac{1}{2} \sum_{p,q=1}^{3} (\beta_p\gamma_q + \gamma_p\beta_q) a_p(x) a_q(x) + \sum_{p=1}^{3} \alpha_p a_p(x),$$

$$g(x) = \sum_{p,q=1}^{3} \beta_{p,q} a_p(x) a_q(x) + \sum_{p=1}^{3} \beta_p a_p(x),$$

$$h(x) = \sum_{p,q=1}^{3} \gamma_{p,q} a_p(x) a_q(x) + \sum_{p=1}^{3} \gamma_p a_p(x),$$

where $\beta_{p,q}, \gamma_{p,q}$ and $\beta_{p,q}\gamma_r + \gamma_{p,q}\beta_r$ are symmetric in p, q, r and

$$\beta_{p,q}\gamma_{s,t} + \gamma_{p,q}\beta_{s,t} = 0 \qquad (p, q, s, t = 1, 2, 3).$$

In the case (10.5b) we have

$$f(x) = \frac{1}{3} \sum_{p,q,r=1}^{3} (\beta_{p,q}\beta_r + \gamma_{p,q}\gamma_r) a_p(x) a_q(x) a_r(x) +$$

$$+ \frac{1}{2} \sum_{p,q=1}^{3} (\beta_p \beta_q + \gamma_p \gamma_q) a_p(x) a_q(x) + \sum_{p=1}^{3} \alpha_p a_p(x),$$

$$g(x) = \sum_{p,q=1}^{3} \beta_{p,q} a_p(x) a_q(x) + \sum_{p=1}^{3} \beta_p a_p(x),$$

$$h(x) = \sum_{p,q=1m}^{3} \gamma_{p,q} a_p(x) a_q(x) + \sum_{p=1}^{3} \gamma_p a_p(x),$$

where $\beta_{p,q}, \gamma_{p,q}$ and $\beta_{p,q}\beta_r + \gamma_{p,q}\gamma_r$ are symmetric in p, q, r and

$$\beta_{p,q}\beta_{s,t} + \gamma_{p,q}\gamma_{s,t} = 0 \qquad (p, q, s, t = 1, 2, 3).$$

In the case (ii) $sp(V) = \{m_1, m_2\}$ with $mult(m_1) = 3$, $mult(m_2) = 1$, hence the general nondegenerate solution is

$$f(x) = \sum_{p,q=1}^{2} \alpha_{p,q} a_p(x) a_q(x) m_1(x) + \sum_{p=1}^{2} \alpha_p a_p(x) m_1(x) + \alpha m_1(x) + \lambda m_2(x),$$

$$g_i(x) = \sum_{p,q=1}^{2} \beta_{p,q}^{(i)} a_p(x) a_q(x) m_1(x) + \sum_{p=1}^{2} \beta_p^{(i)} a_p(x) m_1(x) +$$

$$+ \beta^{(i)} m_1(x) + \mu^{(i)} m_2(x),$$

$$h_i(x) = \sum_{p,q=1}^{2} \gamma_{p,q}^{(i)} a_p(x) a_q(x) m_1(x) + \sum_{p=1}^{2} \gamma_p^{(i)} a_p(x) m_1(x) +$$

$$+\gamma^{(i)}m_1(x) + \nu^{(i)}m_2(x),$$

$(i = 1, 2, 3, 4)$ where m_1, m_2 are different nonzero complex exponentials, a_1, a_2 are linearly independent complex additive functions and all constants are symmetric in the subscripts and are subjected to the conditions

$$
\begin{pmatrix}
\beta_{p,q}^{(1)} & \beta_{p,q}^{(2)} & \beta_{p,q}^{(3)} & \beta_{p,q}^{(4)} \\
\beta_{p}^{(1)} & \beta_{p}^{(2)} & \beta_{p}^{(3)} & \beta_{p}^{(4)} \\
\beta^{(1)} & \beta^{(2)} & \beta^{(3)} & \beta^{(4)} \\
\mu^{(1)} & \mu^{(2)} & \mu^{(3)} & \mu^{(4)}
\end{pmatrix}
\begin{pmatrix}
\gamma_{s,t}^{(1)} & \gamma_{s}^{(1)} & \gamma^{(1)} & \mu^{(1)} \\
\gamma_{s,t}^{(2)} & \gamma_{s}^{(2)} & \gamma^{(2)} & \mu^{(2)} \\
\gamma_{s,t}^{(3)} & \gamma_{s}^{(3)} & \gamma^{(3)} & \mu^{(3)} \\
\gamma_{s,t}^{(4)} & \gamma_{s}^{(4)} & \gamma^{(4)} & \mu^{(4)}
\end{pmatrix}
$$

$$
=
\begin{pmatrix}
0 & 0 & \alpha_{p,q} & 0 \\
0 & 2\alpha_{s,p} & \alpha_p & 0 \\
\alpha_{s,t} & \alpha_s & \alpha & 0 \\
0 & 0 & 0 & \lambda
\end{pmatrix}
$$

for all $p, q, r, s, t = 1, 2$.

In the case (10.5a) we have either $m_1 = 1$ and $m_2 = m \neq 1$, hence

$$f(x) = \frac{1}{2}\sum_{p,q=1}^{2}(\beta_p\gamma_q + \gamma_p\beta_q)a_p(x)a_q(x) + \sum_{p=1}^{2}\alpha_p a_p(x) - 2\beta\gamma + 2\beta\gamma m(x),$$

$$g(x) = \sum_{p=1}^{2}\beta_p a_p(x) + \beta - \beta m(x),$$

$$h(x) = \sum_{p=1}^{2}\gamma_p a_p(x) + \gamma - \gamma m(x),$$

where $\beta_p\gamma + \gamma_p\beta = 0$ $(p = 1, 2)$, or $m_1 \neq 1$ and $m_2 = 1$, and we have no nondegenerate solutions. In the case (10.5b) we have either $m_1 = 1$ and $m_2 = m \neq 1$, hence

$$f(x) = \frac{1}{2}\sum_{p,q=1}^{2}(\beta_p\beta_q + \gamma_p\gamma_q)a_p(x)a_q(x) + \sum_{p=1}^{2}\alpha_p a_p(x) - \beta^2 - \gamma^2 + (\beta^2 + \gamma^2)m(x),$$

$$g(x) = \sum_{p=1}^{2}\beta_p a_p(x) + \beta - \beta m(x),$$

$$h(x) = \sum_{p=1}^{2} \gamma_p a_p(x) + \gamma - \gamma m(x),$$

where $\beta_p \beta + \gamma_p \gamma = 0$ $(p = 1, 2)$, or $m_1 \neq 1$ and $m_2 = 1$, and we have no nondegenerate solutions.

In the case (iii) $sp(V) = \{m_1, m_2\}$ with $mult(m_1) = mult(m_2) = 2$, hence the general nondegenerate solution is

$$f(x) = (\alpha_1 a_1(x) + \alpha_2) m_1(x) + (\alpha_3 a_2(x) + \alpha_4) m_2(x),$$

$$g_i(x) = (\beta_1^{(i)} a_1(x) + \beta_2^{(i)}) m_1(x) + (\beta_3^{(i)} a_2(x) + \beta_4^{(i)}) m_2(x),$$

$$h_i(x) = (\gamma_1^{(i)} a_1(x) + \gamma_2^{(i)}) m_1(x) + (\gamma_3^{(i)} a_2(x) + \gamma_4^{(i)}) m_2(x),$$

$(i = 1, 2, 3, 4)$ where m_1, m_2 are different nonzero complex exponentials, a_1, a_2 are nonzero complex additive functions and the constants are subjected to the conditions

$$\begin{pmatrix} \beta_1^{(1)} & \beta_1^{(2)} & \beta_1^{(3)} & \beta_1^{(4)} \\ \beta_2^{(1)} & \beta_2^{(2)} & \beta_2^{(3)} & \beta_2^{(4)} \\ \beta_3^{(1)} & \beta_3^{(2)} & \beta_3^{(3)} & \beta_3^{(4)} \\ \beta_4^{(1)} & \beta_4^{(2)} & \beta_4^{(3)} & \beta_4^{(4)} \end{pmatrix} \begin{pmatrix} \gamma_1^{(1)} & \gamma_2^{(1)} & \gamma_3^{(1)} & \gamma_4^{(1)} \\ \gamma_1^{(2)} & \gamma_2^{(2)} & \gamma_3^{(2)} & \gamma_4^{(2)} \\ \gamma_1^{(3)} & \gamma_2^{(3)} & \gamma_3^{(3)} & \gamma_4^{(3)} \\ \gamma_1^{(4)} & \gamma_2^{(4)} & \gamma_3^{(4)} & \gamma_4^{(4)} \end{pmatrix} =$$

$$= \begin{pmatrix} 0 & \alpha_1 & 0 & 0 \\ \alpha_1 & \alpha_2 & 0 & 0 \\ 0 & 0 & 0 & \alpha_3 \\ 0 & 0 & \alpha_3 & \alpha_4 \end{pmatrix}.$$

In the case (10.5a) we have $m_1 = 1$, $m_2 = m \neq 1$ and

$$f(x) = a_1(x) - 2\mu\nu - (\mu\gamma + \nu\beta) a_2(x) m(x) + 2\mu\nu m(x),$$

$$g(x) = \mu + \beta a_2 m(x) - \mu m(x), \qquad h(x) = \nu + \gamma a_2 m(x) - \nu m(x),$$

where $\beta\gamma = 0$. In the case (10.5b) we have $m_1 = 1$, $m_2 = m \neq 1$ and

$$f(x) = a_1(x) - \mu^2 - \nu^2 - (\mu + i\nu)\beta a_2(x) m(x) + (\mu^2 + \nu^2) m(x),$$

$$g(x) = \mu + \beta a_2(x) m(x) - \mu m(x), \qquad h(x) = \nu + i\beta a_2(x) m(x) - \nu m(x).$$

In the case (iv) $sp(V) = \{m_1, m_2, m_3\}$ with $mult(m_1) = 2$, $mult(m_2) = mult(m_3) = 1$ hence the general nondegenerate solution is

$$f(x) = (\alpha_1 a(x) + \alpha_2) m_1(x) + \alpha_3 m_2(x) + \alpha_4 m_3(x),$$

$$g_i(x) = (\beta_1^{(i)}a(x) + \beta_2^{(i)})m_1(x) + \beta_3^{(i)}m_2(x) + \beta_4^{(i)}m_3(x),$$

$$h_i(x) = (\gamma_1^{(i)}a(x) + \gamma_2^{(i)})m_1(x) + \gamma_3^{(i)}m_2(x) + \gamma_4^{(i)}m_3(x),$$

$(i = 1, 2, 3, 4)$ where m_1, m_2, m_3 are different nonzero complex exponentials, a is a nonzero complex additive function and the constants are subjected to the conditions

$$
\begin{pmatrix}
\beta_1^{(1)} & \beta_1^{(2)} & \beta_1^{(3)} & \beta_1^{(4)} \\
\beta_2^{(1)} & \beta_2^{(2)} & \beta_2^{(3)} & \beta_2^{(4)} \\
\beta_3^{(1)} & \beta_3^{(2)} & \beta_3^{(3)} & \beta_3^{(4)} \\
\beta_4^{(1)} & \beta_4^{(2)} & \beta_4^{(3)} & \beta_4^{(4)}
\end{pmatrix}
\begin{pmatrix}
\gamma_1^{(1)} & \gamma_2^{(1)} & \gamma_3^{(1)} & \gamma_4^{(1)} \\
\gamma_1^{(2)} & \gamma_2^{(2)} & \gamma_3^{(2)} & \gamma_4^{(2)} \\
\gamma_1^{(3)} & \gamma_2^{(3)} & \gamma_3^{(3)} & \gamma_4^{(3)} \\
\gamma_1^{(4)} & \gamma_2^{(4)} & \gamma_3^{(4)} & \gamma_4^{(4)}
\end{pmatrix}
=
$$

$$
=
\begin{pmatrix}
0 & \alpha_1 & 0 & 0 \\
\alpha_1 & \alpha_2 & 0 & 0 \\
0 & 0 & \alpha_3 & 0 \\
0 & 0 & 0 & \alpha_4
\end{pmatrix}.
$$

Hence in the case (10.5a) we have either $m_1 = 1$, $m_2 \neq 1$, $m_3 \neq 1$ and

$$f(x) = a(x) - 2\beta_1\gamma_1 - 2\beta_2\gamma_2 + 2\beta_1\gamma_1 m_2(x) + 2\beta_2\gamma_2 m_3(x),$$

$$g(x) = -(\beta_1 + \beta_2) + \beta_1 m_2(x) + \beta_2 m_3(x),$$

$$h(x) = -(\gamma_1 + \gamma_2) + \gamma_1 m_2(x) + \gamma_2 m_3(x),$$

where $\beta_1\gamma_2 + \gamma_1\beta_2 = 0$, or $m_1 \neq 1$ and we have no nondegenerate solutions, and in the case (10.5b) we have either $m_1 = 1$, $m_2 \neq 1$, $m_3 \neq 1$ and

$$f(x) = a(x) - \beta_1^2 - \beta_2^2 - \gamma_1^2 - \gamma_2^2 + (\beta_1^2 + \gamma_1^2)m_2(x) + (\beta_2^2 + \gamma_2^2)m_3(x),$$

$$g(x) = -(\beta_1 + \beta_2) + \beta_1 m_2(x) + \beta_2 m_3(x),$$

$$h(x) = -(\gamma_1 + \gamma_2) + \gamma_1 m_2(x) + \gamma_2 m_3(x),$$

or $m_1 \neq 1$ and we have no nondegenerate solutions.

In the case (v) $sp(V) = \{m_1, m_2, m_3, m_4\}$ with $mult(m_i) = 1$ ($i = 1, 2, 3, 4$) and the general nondegenerate solution is

$$f(x) = \alpha_1 m_1(x) + \alpha_2 m_2(x) + \alpha_3 m_3(x) + \alpha_4 m_4(x),$$

$$g_i(x) = \beta_1^{(i)} m_1(x) + \beta_2^{(i)} m_2(x) + \beta_3^{(i)} m_3(x) + \beta_4^{(i)} m_4(x),$$

$$h_i(x) = \gamma_1^{(i)} m_1(x) + \gamma_2^{(i)} m_2(x) + \gamma_3^{(i)} m_3(x) + \gamma_4^{(i)} m_4(x),$$

$(i = 1, 2, 3, 4)$ where m_1, m_2, m_3, m_4 are different nonzero complex exponentials, and the constants are subjected to the conditions

$$\begin{pmatrix} \beta_1^{(1)} & \beta_1^{(2)} & \beta_1^{(3)} & \beta_1^{(4)} \\ \beta_2^{(1)} & \beta_2^{(2)} & \beta_2^{(3)} & \beta_2^{(4)} \\ \beta_3^{(1)} & \beta_3^{(2)} & \beta_3^{(3)} & \beta_3^{(4)} \\ \beta_4^{(1)} & \beta_4^{(2)} & \beta_4^{(3)} & \beta_4^{(4)} \end{pmatrix} \begin{pmatrix} \gamma_1^{(1)} & \gamma_2^{(1)} & \gamma_3^{(1)} & \gamma_4^{(1)} \\ \gamma_1^{(2)} & \gamma_2^{(2)} & \gamma_3^{(2)} & \gamma_4^{(2)} \\ \gamma_1^{(3)} & \gamma_2^{(3)} & \gamma_3^{(3)} & \gamma_4^{(3)} \\ \gamma_1^{(4)} & \gamma_2^{(4)} & \gamma_3^{(4)} & \gamma_4^{(4)} \end{pmatrix} =$$

$$= \begin{pmatrix} \alpha_1 & 0 & 0 & 0 \\ 0 & \alpha_2 & 0 & 0 \\ 0 & 0 & \alpha_3 & 0 \\ 0 & 0 & 0 & \alpha_4 \end{pmatrix}.$$

In the case (10.5a) and (10.5b) we have no nondegenerate solutions.

The following example is the functional equation

$$(10.6) \qquad f(x+y) = g_1(x)h_1(y) + g_2(x)h_2(y) + g_3(x)h_3(y),$$

where we suppose that the complex valued functions $f, g_1, g_2, g_3, h_1, h_2, h_3$ form a nondegenerate solution. In particular, we consider the special cases

$$f = g_1 = h_2, \qquad g_2 = h_1, \qquad g_3 = h_3,$$

corresponding to the equation

$$(10.6a) \qquad f(x+y) = f(x)g(y) + f(y)g(x) + h(x)h(y),$$

and

$$h_1 = g_2 = 1,$$

corresponding to the equation

$$(10.6b) \qquad f(x+y) = g(x) + h(y) + k(x)l(y).$$

In the special case (10.6b) the spectrum of the equation contains the exponential 1. In general, for the spectrum of (10.6) we have three different possibilities, corresponding to the following decompositions of the number 3:

(i) $3 = 3$
(ii) $3 = 2 + 1$

89

(iii) $3 = 1 + 1 + 1$.

In the case (i) $sp(V) = \{m\}$ with $mult(m) = 3$ and the general nondegenerate solution is

$$f(x) = \sum_{p,q=1}^{2} \alpha_{p,q} a_p(x) a_q(x) m(x) + \sum_{p=1}^{2} \alpha_p a_p(x) m(x) + \alpha m(x),$$

$$g_i(x) = \sum_{p,q=1}^{2} \beta_{p,q}^{(i)} a_p(x) a_q(x) m(x) + \sum_{p=1}^{2} \beta_p^{(i)} a_p(x) m(x) + \beta^{(i)} m(x),$$

$$h_i(x) = \sum_{p,q=1}^{2} \gamma_{p,q}^{(i)} a_p(x) a_q(x) m(x) + \sum_{p=1}^{2} \gamma_p^{(i)} a_p(x) m(x) + \gamma^{(i)} m(x),$$

($i = 1, 2, 3$) where m is a nonzero complex exponential, a_1, a_2 are linearly independent complex additive functions and the constants are all symmetric in the subscripts and are subjected to the conditions

$$\begin{pmatrix} \beta_{p,q}^{(1)} & \beta_{p,q}^{(2)} & \beta_{p,q}^{(3)} \\ \beta_p^{(1)} & \beta_p^{(2)} & \beta_p^{(3)} \\ \beta^{(1)} & \beta^{(2)} & \beta^{(3)} \end{pmatrix} \begin{pmatrix} \gamma_{s,t}^{(1)} & \gamma_s^{(1)} & \gamma^{(1)} \\ \gamma_{s,t}^{(2)} & \gamma_s^{(2)} & \gamma^{(2)} \\ \gamma_{s,t}^{(3)} & \gamma_s^{(3)} & \gamma^{(3)} \end{pmatrix} = \begin{pmatrix} 0 & 0 & \alpha_{p,q} \\ 0 & 2\alpha_{s,p} & \alpha_p \\ \alpha_{s,t} & \alpha_s & \alpha \end{pmatrix}$$

for all $p, q, s, t = 1, 2$.

In the case $(10.6a)$ we have

$$f(x) = \sum_{p,q=1}^{2} \alpha_{p,q} a_p(x) a_q(x) m(x) + \sum_{p=1}^{2} \alpha_p a_p(x) m(x) + \alpha m(x),$$

$$g(x) = \sum_{p,q=1}^{2} \beta_{p,q} a_p(x) a_q(x) m(x) + \sum_{p=1}^{2} \beta_p a_p(x) m(x) + \beta m(x),$$

$$h(x) = \sum_{p,q=1}^{2} \gamma_{p,q} a_p(x) a_q(x) m(x) + \sum_{p=1}^{2} \gamma_p a_p(x) m(x) + \gamma m(x),$$

where $\alpha_{p,q}, \beta_{p,q}, \gamma_{p,q}$ are symmetric in p, q and

$$\begin{pmatrix} \alpha_{p,q} & \beta_{p,q} & \gamma_{p,q} \\ \alpha_p & \beta_p & \gamma_p \\ \alpha & \beta & \gamma \end{pmatrix} \begin{pmatrix} \beta_{s,t} & \beta_s & \beta \\ \alpha_{s,t} & \alpha_s & \alpha \\ \gamma_{s,t} & \gamma_s & \gamma \end{pmatrix} = \begin{pmatrix} 0 & 0 & \alpha_{p,q} \\ 0 & 2\alpha_{s,p} & \alpha_p \\ \alpha_{s,t} & \alpha_s & \alpha \end{pmatrix}$$

holds for all $p, q, s, t = 1, 2$.

In the case $(10.6b)$ we have

$$f(x) = \frac{1}{2} \sum_{p,q=1}^{2} \delta_p \epsilon_q a_p(x) a_q(x) + \sum_{p=1}^{2} (\gamma_p + \delta \epsilon_p) a_p(x) + \beta + \gamma + \delta \epsilon,$$

$$g(x) = \frac{1}{2} \sum_{p,q=1}^{2} \delta_p \epsilon_q a_p(x) a_q(x) + \sum_{p=1}^{2} \beta_p a_p(x) + \beta,$$

$$h(x) = \frac{1}{2} \sum_{p,q=1}^{2} \delta_p \epsilon_q a_p(x) a_q(x) + \sum_{p=1}^{2} \gamma_p a_p(x) + \gamma,$$

$$k(x) = \sum_{p=1}^{2} \delta_p a_p(x) + \delta, \qquad l(x) = \sum_{p=1}^{2} \epsilon_p a_p(x) + \epsilon,$$

where $\beta_p + \epsilon \delta_p = \gamma_p + \delta \epsilon_p$ $(p = 1, 2)$ and $\delta_1 \epsilon_2 = \delta_2 \epsilon_1$.

In the case (ii) $sp(V) = \{m_1, m_2\}$ with $mult(m_1) = 2$, $mult(m_2) = 1$ and the general nondegenerate solution is

$$f(x) = (\alpha_1 a(x) + \alpha_2) m_1(x) + \alpha_3 m_2(x),$$

$$g_i(x) = (\beta_1^{(i)} a(x) + \beta_2^{(i)}) m_1(x) + \beta_3^{(i)} m_2(x),$$

$$h_i(x) = (\gamma_1^{(i)} a(x) + \gamma_2^{(i)}) m_1(x) + \gamma_3^{(i)} m_2(x),$$

$(i = 1, 2, 3)$ where m_1, m_2 are different nonzero complex exponentials, a is a nonzero complex additive function and the constants are subjected to the conditions

$$\begin{pmatrix} \beta_1(1) & \beta_1(2) & \beta_1(3) \\ \beta_2(1) & \beta_2(2) & \beta_2(3) \\ \beta_3(1) & \beta_3(2) & \beta_3(3) \end{pmatrix} \begin{pmatrix} \gamma_1^{(1)} & \gamma_2^{(1)} & \gamma_3^{(1)} \\ \gamma_1^{(2)} & \gamma_2^{(2)} & \gamma_3^{(2)} \\ \gamma_1^{(3)} & \gamma_2^{(3)} & \gamma_3^{(3)} \end{pmatrix} = \begin{pmatrix} 0 & \alpha_1 & 0 \\ \alpha_1 & \alpha_2 & 0 \\ 0 & 0 & \alpha_3 \end{pmatrix}.$$

In the case $(10.6a)$

$$f(x) = (\alpha_1 a(x) + \alpha_2) m_1(x) + \alpha_3 m_2(x),$$

$$g(x) = (\beta_1 a(x) + \beta_2) m_1(x) + \beta_3 m_2(x),$$

$$h(x) = (\gamma_1 a(x) + \gamma_2) m_1(x) + \gamma_3 m_2(x),$$

91

where

$$\begin{pmatrix} \alpha_1 & \beta_1 & \gamma_1 \\ \alpha_2 & \beta_2 & \gamma_2 \\ \alpha_3 & \beta_3 & \gamma_3 \end{pmatrix} \begin{pmatrix} \beta_1 & \beta_2 & \beta_3 \\ \alpha_1 & \alpha_2 & \alpha_3 \\ \gamma_1 & \gamma_2 & \gamma_3 \end{pmatrix} = \begin{pmatrix} 0 & \alpha_1 & 0 \\ \alpha_1 & \alpha_2 & 0 \\ 0 & 0 & \alpha_3 \end{pmatrix}.$$

In the case (10.6b) either $m_1 = 1$ and $m_2 = m \neq 1$ hence

$$f(x) = \alpha_1 a(x) + \beta_2 + \delta_2 + \gamma_2 \epsilon_2 + \gamma_3 \epsilon_3 m(x),$$

$$g(x) = \alpha_1 a(x) + \beta_2 - \gamma_3 \epsilon_2 m(x), \qquad h(x) = \alpha_1 a(x) + \delta_2 - \gamma_2 \epsilon_3 m(x),$$

$$k(x) = \gamma_2 + \gamma_3 m(x), \qquad l(x) = \epsilon_2 + \epsilon_3 m(x);$$

or $m_1 \neq 1$, $m_2 = 1$ and we have no nondegenerate solution.

Finally, in the case (iii) $sp(V) = \{m_1, m_2, m_3\}$ with $mult(m_1) = = mult(m_2) = mult(m_3) = 1$ and the general nondegenerate solution is

$$f(x) = \alpha_1 m_1(x) + \alpha_2 m_2(x) + \alpha_3 m_3(x),$$

$$g_i(x) = \beta_1^{(i)} m_1(x) + \beta_2^{(i)} m_2(x) + \beta_3^{(i)} m_3(x),$$

$$h_i(x) = \gamma_1^{(i)} m_1(x) + \gamma_2^{(i)} m_2(x) + \gamma_3^{(i)} m_3(x),$$

($i = 1, 2, 3$) where m_1, m_2, m_3 are different nonzero complex exponentials, and the constants are subjected to the conditions

$$\begin{pmatrix} \beta_1^{(1)} & \beta_1^{(2)} & \beta_1^{(3)} \\ \beta_2^{(1)} & \beta_2^{(2)} & \beta_2^{(3)} \\ \beta_3^{(1)} & \beta_3^{(2)} & \beta_3^{(3)} \end{pmatrix} \begin{pmatrix} \gamma_1^{(1)} & \gamma_2^{(1)} & \gamma_3^{(1)} \\ \gamma_1^{(2)} & \gamma_2^{(2)} & \gamma_3^{(2)} \\ \gamma_1^{(3)} & \gamma_2^{(3)} & \gamma_3^{(3)} \end{pmatrix} = \begin{pmatrix} \alpha_1 & 0 & 0 \\ 0 & \alpha_2 & 0 \\ 0 & 0 & \alpha_3 \end{pmatrix}.$$

In the case (10.6a)

$$f(x) = \alpha_1 m_1(x) + \alpha_2 m_2(x) + \alpha_3 m_3(x),$$

$$g(x) = \beta_1 m_1(x) + \beta_2 m_2(x) + \beta_3 m_3(x),$$

$$h(x) = \gamma_1 m_1(x) + \gamma_2 m_2(x) + \gamma_3 m_3(x),$$

with

$$\begin{pmatrix} \alpha_1 & \beta_1 & \gamma_1 \\ \alpha_2 & \beta_2 & \gamma_2 \\ \alpha_3 & \beta_3 & \gamma_3 \end{pmatrix} \begin{pmatrix} \beta_1 & \beta_2 & \beta_3 \\ \alpha_1 & \alpha_2 & \alpha_3 \\ \gamma_1 & \gamma_2 & \gamma_3 \end{pmatrix} = \begin{pmatrix} \alpha_1 & 0 & 0 \\ 0 & \alpha_2 & 0 \\ 0 & 0 & \alpha_3 \end{pmatrix},$$

and in the case (10.6b) we have no nondegenerate solutions.

Now we consider the case $n = 2$, that is, the functional equation

(10.7) $$f(x + y) = g_1(x)h_1(y) + g_2(x)h_2(y),$$

where the complex valued functions f, g_1, g_2, h_1, h_2 form a nondegenerate solution. In particular, we consider the special cases

$$f = g_1 = h_2, \qquad g = g_2 = h_1,$$

corresponding to the sine-equation

(10.7a) $$f(x + y) = f(x)g(y) + f(y)g(x),$$

and

$$f = g_1 = h_1, \qquad g = g_2 = -h_2,$$

corresponding to the cosine-equation

(10.7b) $$f(x + y) = f(x)f(y) - g(x)g(y).$$

For the spectrum of (10.7) we have two possibilities:
(i) $2 = 2$
(ii) $2 = 1 + 1$.

In the case (i) $sp(V) = \{m\}$ with $mult(m) = 2$ and the general nondegenerate solution is

$$f(x) = (\alpha_1 a(x) + \alpha_2)m(x),$$

$$g_i(x) = (\beta_1^{(i)} a(x) + \beta_2^{(i)})m(x),$$

$$h_i(x) = (\gamma_1^{(i)} a(x) + \gamma_2^{(i)})m(x),$$

$(i = 1, 2)$, where m is a nonzero complex exponential, a is a nonzero complex additive function and the constants are subjected to the condition

$$\begin{pmatrix} \beta_1^{(1)} & \beta_1^{(2)} \\ \beta_2^{(1)} & \beta_2^{(2)} \end{pmatrix} \begin{pmatrix} \gamma_1^{(1)} & \gamma_2^{(1)} \\ \gamma_1^{(2)} & \gamma_2^{(2)} \end{pmatrix} = \begin{pmatrix} 0 & \alpha_1 \\ \alpha_1 & \alpha_2 \end{pmatrix}.$$

In the case (10.7a) we have

$$f(x) = a(x)m(x),$$

$$g(x) = m(x),$$

93

and in the case (10.7b) we have

$$f(x) = (\pm ia(x) + 1)m(x),$$

$$g(x) = a(x)m(x).$$

In the case (ii) $sp(V) = \{m_1, m_2\}$ with $mult(m_1) = mult(m_2) = 1$ and the general nondegenerate solution is

$$f(x) = \alpha_1 m_1(x) + \alpha_2 m_2(x),$$

$$g_i(x) = \beta_1^{(i)} m_1(x) + \beta_2^{(i)} m_2(x),$$

$$h_i(x) = \gamma_1^{(i)} m_1(x) + \gamma_2^{(i)} m_2(x),$$

($i = 1, 2$) where m_1, m_2 are different nonzero complex exponentials, and the constants are subjected to the condition

$$\begin{pmatrix} \beta_1^{(1)} & \beta_1^{(2)} \\ \beta_2^{(1)} & \beta_2^{(2)} \end{pmatrix} \begin{pmatrix} \gamma_1^{(1)} & \gamma_2^{(1)} \\ \gamma_1^{(2)} & \gamma_2^{(2)} \end{pmatrix} = \begin{pmatrix} \alpha_1 & 0 \\ 0 & \alpha_2 \end{pmatrix}.$$

In the case (10.7a) we have

$$f(x) = \alpha(m_1(x) - m_2(x)),$$

$$g(x) = \frac{1}{2}(m_1(x) + m_2(x)),$$

and in the case (10.7b) we have

$$f(x) = \alpha_1 m_1(x) + \alpha_2 m_2(x),$$

$$g(x) = \beta_1 m_1(x) + \beta_2 m_2(x),$$

with

$$\alpha_1^2 + \beta_1^2 = \alpha_1,$$

$$\alpha_1 \alpha_2 + \beta_1 \beta_2 = 0,$$

$$\alpha_2^2 + \beta_2^2 = \alpha_2.$$

Finally, in the trivial case $n = 1$, that is, in the case of the Pexider-equation

$$f(x + y) = g_1(x)h_1(y)$$

94

we have the only nondegenerate solution

$$f(x) = \alpha\beta m(x),$$

$$g_1(x) = \alpha m(x),$$

$$h_1(x) = \beta m(x),$$

where m is a nonzero complex exponential and α, β are constants.

If $n > 4$ then there are more subcases to consider but the method works similarly.

We note, that the above method works also on nondiscrete topological abelian groups, if the functions f, g_i, h_i are continuous functions. In this case all the exponentials and additive functions in the representation (10.2) are continuous. Using the results of Section 3 and Section 5 the condition on the continuity of f, g_i, h_i can be weakened.

We noted that the fundamental Theorem 10.1 can be proved also for commutative topological semigroups by using the method of [MCK6], [MCK7] (see also [SZÉ9]) and even for commutative groupoids. Also the range of the unknown functions can be more general. For the details see [MCK7].

REFERENCES 10.6.

For further results concerning (10.1) and its special cases under various assumptions can be found in [ACZ1], [ACZ2], [ACZ3], [ACZ4], [ACZ5], [ACZ6], [ACZ7], [ACZ9], [ACZ10], [ACZ13], [ACZ14], [ACZ15], [ACZ16], [ACZ18], [ANS], [BEL], [ENG], [GHE3], [HAJ], [ILS1], [ILS2], [ILS3], [KCZ1], [KCZ2], [KCZ3], [KCZ4], [KEM1], [KEM4], [LAI2], [LES1], [LES2], [LEV], [MAG], [MCK2], [MCK5], [MCK6], [MCK7], [NOV], [OCO1], [OCO2], [OCO3], [PER], [PGA2], [REI1], [REI2], [ROT], [SAK], [SAT], [SCH], [SCW], [STA], [STE], [SZÉ4], [SZÉ6], [SZÉ7], [SZÉ9], [SZÉ11], [SZÉ15], [SZÉ19], [SZÉ24], [UNG1], [UNG2], [VIN1], [VIN2], [VIN3].

11. D'Alembert-type functional equations

This section is devoted to the study of the functional equation

$$(11.1) \qquad f(x+y) + g(x-y) = \sum_{i=1}^{n} h_i(x)k_i(y)$$

on commutative abelian groups. Equation (11.1) is obviously a generalization of the Levi-Civitá-equation (10.1), but it also generalizes e.g. the square-norm equation (9.6), and the d'Alembert-equation

$$(11.2) \qquad f(x+y) + f(x-y) = 2f(x)f(y).$$

Equations of the form (11.1) have been treated by many authors. In [SAT] (see also [ACZ7], [KEM1]) regular solutions of (11.1) are determined by reducing it to differential equations. In [ACZ6] the open problem about the general solution of (11.1) is formulated. This problem has been solved in [PEN], [RUK1], [RUK2], [SZÉ6], by different methods. Here we present a new method based on spectral synthesis.

In contrast with (10.1) it is not true in general, that all complex valued functions f in (11.1) must be normal exponential polynomials. To see this, it is enough to take any complex valued symmetric and biadditive function $B : \mathbf{R}^2 \to \mathbf{R}$ with the property, that the linear space spanned by the additive functions $x \mapsto B(x, y)$ for all y in $\mathbf{R}$ is not of finite dimension. Then the function $f = B^*$ is a solution of the square-norm equation (9.6), but $\tau(f)$ is of infinite dimension, because it contains all the functions $x \mapsto B(x, y)$ for all y in $\mathbf{R}$. Hence f is not a normal exponential polynomial.

It is easy to see that (11.1) can be reduced to the two equations

$$(11.1a) \qquad \varphi(x+y) + \varphi(x-y) = \sum_{i=1}^{n} h_i(x)k_i(y),$$

$$(11.1b) \qquad \psi(x+y) - \psi(x-y) = \sum_{i=1}^{n} h_i(x)k_i(y),$$

where k_i is even in (11.1a) and is odd in (11.1b), respectively ($i = 1, 2, \ldots n$). Indeed, substituting $-y$ for y in (11.1) and then adding, resp. subtracting the new equation to, resp. from (11.1) we get (11.1a) for $\varphi = f + g$ with $k_i + \check{k}_i$ instead of k_i, resp. (11.1b) for $\psi = f - g$ with $k_i - \check{k}_i$ instead of k_i ($i = 1, 2, \ldots, n$). First we deal with (11.1a).

THEOREM 11.1. *Let G be an abelian group, n a positive integer and φ, h_i, k_i : $G \to \mathbf{C}$ functions ($i = 1, 2, \ldots, n$) satisfying (11.1a) for all x, y in G. Then there exist nonnegative integers k, l with $k + l = n$, nonnegative integers $n_1, n_2, \ldots, n_k, n_{k+1}, \ldots, n_{k+l}$ with*

$$n_1 + n_2 + \cdots + n_k + \left[\frac{n_{k+1} - 1}{2} \right] + \cdots + \left[\frac{n_{k+l} - 1}{2} \right] + l = n,$$

further there exist different nonzero complex exponentials $m_1, m_2, \ldots, m_k$, $m_{k+1}, \ldots, m_{k+l}$ with $m_i \neq m_j$ for $i, j = 1, 2, \ldots, k$ and $m_{k+i} = \check{m}_{k+i}$ for $i = 1, 2, \ldots, l$, and there exist normal polynomials $p_i, q_i : G \to \mathbf{C}$ for $i = 1, 2, \ldots, k$ and polynomials $r_j : G \to \mathbf{C}$ for $j = 1, 2, \ldots, l$ with $\deg p_i, \deg q_i \leq n_i - 1$ and $\deg r_j \leq n_{k+j} - 1$ for $i = 1, 2, \ldots, k; j = 1, 2, \ldots, l$, such that

$$\varphi(x) = \sum_{i=1}^{k} \left(p_i(x) m_i(x) + q_i(x) \check{m}_i(x) \right) + \sum_{j=1}^{l} r_j(x) m_{k+j}(x)$$

holds for all x, y in G. If $\{k_1, k_2, \ldots, k_n\}$, resp. $\{h_1, h_2, \ldots, h_n\}$ are linearly independent, then $\{h_1, h_2, \ldots, h_n\}$, resp. $\{k_1, k_2, \ldots, k_n\}$ have the same form. (Here $[x]$ denotes the integer part of x.)

PROOF: In order to apply spectral synthesis by the results in Section 8 we may suppose that G is finitely generated. If all the functions $k_1, k_2, \ldots, k_n$ are zero, then (11.1a) implies immediately, that φ is zero. Hence -by reducing n if necessary - we may suppose, that $\{k_1, k_2, \ldots, k_n\}$ are linearly independent. It follows that there are constants $\lambda_{i,j}$ in $\mathbf{C}$ and elements y_j in G ($i, j = 1, 2, \ldots, n$) with

$$(11.2) \qquad h_i(x) = \sum_{j=1}^{n} \lambda_{i,j} \big(\varphi(x + y_j) + \varphi(x - y_j) \big)$$

for x in G and $i = 1, 2, \ldots, n$. Substitution into (11.1a) gives the functional equation

$$(11.3) \qquad \left(\delta_y + \delta_{-y} - \sum_{i=1}^{n} \sum_{j=1}^{n} \lambda_{i,j} k_i(y) (\delta_{y_j} + \delta_{y_{-j}}) \right) * \varphi = 0$$

for all y in G. (Here δ_y denotes the discrete measure concentrated at the point y.) The solution space V of (11.3) is obviously a closed translation invariant linear space, that is, a variety in $\mathcal{C}(G)$. If φ is not zero, then this variety contains nonzero exponential monomials. Let $a = (a_1, a_2, \ldots, a_k)$, where $\{a_1, a_2, \ldots, a_k\}$ are linearly independent real additive functions, let $\alpha = (\alpha_1, \alpha_2, \ldots, \alpha_k)$ be a multi-index and let m be a nonzero complex exponential. Suppose that the exponential monomial $a^\alpha m$ is in V. Then substitution into (11.3) and the use of Lemma 2.8 shows that the functions $a^\beta (m + (-1)^{|\beta|} \breve{m})$ belong to the linear hull of $\{k_1, k_2, \ldots, k_n\}$ for all $\beta \leq \alpha$. Now we can use Lemma 4.8 to infer, that $sp(V)$ is finite, and there exists a positive integer N with the property, that any exponential monomial pm in V satisfies $\deg p \leq N$. The rest of the statement follows then from Theorem 4.6.

By the above proof it follows, that there may exist only a finite number of linearly independent real additive functions a with am in V, for any $m \neq \breve{m}$ in V, hence the above representation of φ can be rewritten in the form

$$\varphi(x) = \sum_{i=1}^{k} \big(P_i(a_{i,1}(x), a_{i,2}(x), \ldots, a_{i,n_i-1}(x)) m_i(x) +$$

$$+ Q_i(a_{i,1}(x), a_{i,2}(x), \ldots, a_{i,n_i-1}(x)) m_i(-x) \big) + \sum_{j=1}^{l} r_j(x) m_{k+j}(x)$$

for all x in G. Here the functions $a_{i,1}, a_{i,2}, \ldots, a_{i,n_i-1}$ are linearly independent real additive functions, further $P_i, Q_i : \mathbf{C}^{n_i-1} \to \mathbf{C}$ $(i = 1, 2, \ldots, k)$ are complex polynomials in $n_i - 1$ complex variables and of degree at most $n_i - 1$ for every $i = 1, 2, \ldots, k$. The meaning of the other parameters is the same as above.

If the surjectivity of multiplication by 2 in G has been supposed, then the only nonzero complex exponential m with the property $m = \breve{m}$ is $m = 1$, hence the last term in the above representation is simply a polynomial.

A modification of the above proof gives the respective theorem for equation (11.1b).

THEOREM 11.2. *Let G be an abelian group, n a positive integer and $\psi, h_i, k_i :$ $G \to \mathbf{C}$ functions $(i = 1, 2, \ldots, n)$ satisfying (11.1b) for all x, y in G. Then there exists a function $F : G/2G \to \mathbf{C}$, there exist nonnegative integers k, l with $k + l = n$, nonnegative integers $n_1, n_2, \ldots, n_k, n_{k+1}, \ldots, n_{k+l}$ with*

$$n_1 + n_2 + \cdots + n_k + \left[\frac{n_{k+1}}{2} \right] + \cdots + \left[\frac{n_{k+l}}{2} \right] = n,$$

further there exist different nonzero complex exponentials $m_1, m_2, \ldots, m_k$, $m_{k+1}, \ldots, m_{k+l}$ with $m_i \neq \check{m}_j$ for $i, j = 1, 2, \ldots, k$ and $m_{k+i} = \check{m}_{k+i}$ for $i = 1, 2, \ldots, l$, and there exist normal polynomials $p_i, q_i : G \to \mathbf{C}$ for $i = 1, 2, \ldots, k$ and polynomials $r_j : G \to \mathbf{C}$ for $j = 1, 2, \ldots, l$ with $\deg p_i, \deg q_i \leq n_i - 1$ and $\deg r_j \leq n_{k+j} - 1$ for $i = 1, 2, \ldots, k; j = 1, 2, \ldots, l$ such that

$$\varphi(x) = \sum_{i=1}^{k} (p_i(x) m_i(x) + q_i(x) \check{m}_i(x)) + \sum_{j=1}^{l} r_j(x) m_{k+j}(x) + F(\Phi(x))$$

holds for all x, y in G, where $\Phi : G \to G/2G$ denotes the natural homomorphism. If $\{k_1, k_2, \ldots, k_n\}$, resp. $\{h_1, h_2, \ldots, h_n\}$ are linearly independent, then $\{h_1, h_2, \ldots, h_n\}$, resp. $\{k_1, k_2, \ldots, k_n\}$ have the same form.

PROOF: If all the functions $k_1, k_2, \ldots, k_n$ are zero, then (11.1b) implies immediately, that $\varphi = F \circ \Phi$ with some function $F : G/2G \to \mathbf{C}$. Hence -by reducing n if necessary - we may suppose, that $\{k_1, k_2, \ldots, k_n\}$ are linearly independent. Using a similar argument as in Theorem 11.1 we get that any exponential polynomial in $\tau(\psi)$ has the form

$$\epsilon(x) = \sum_{i=1}^{k} (p_i(x) m_i(x) + q_i(x) \check{m}_i(x)) + \sum_{j=1}^{l} r_j(x) m_{k+j}(x) + E(x),$$

where the degrees of the polynomials p_i, q_i, r_j do not exceed some fixed positive integer N, and $E : G \to \mathbf{C}$ is a function with the property $E(x + 2y) = E(x)$ for all x, y in G. Obviously we may suppose here, that the polynomials r_j are nonconstant, whenever l is positive. But in this case a slight modification of the proof of Theorem 4.6 shows, that all functions of this type form a closed subspace in $\mathcal{C}(G)$, and then our statement follows.

Similarly as before, we get the following representation for the solution ψ of (11.1b):

$$\psi(x) = \sum_{i=1}^{k} (P_i(a_{i,1}(x), a_{i,2}(x), \ldots, a_{i,n_i-1}(x)) m_i(x) +$$

$$+ Q_i(a_{i,1}(x), a_{i,2}(x), \ldots, a_{i,n_i-1}(x)) \check{m}_i(x)) + \sum_{j=1}^{l} r_j(x) m_{k+j}(x) + F(\Phi(x))$$

with the same meaning of the parameters as above. Obviously, if multiplication by 2 is surjective in G, then F is constant, and the last two terms above form simply a polynomial.

Combining Theorem 11.1 and Theorem 11.2 we get a representation for the solutions f, g of (11.1). The details are left to the reader. We formulate here the following theorem only.

THEOREM 11.3. *Let G be an abelian group in which multiplication by 2 is surjective, n a positive integer and $f, g, h_i, k_i : G \to \mathbf{C}$ functions ($i = 1, 2, \ldots, n$) satisfying (11.1) for all x, y in G. Then f and g are exponential polynomials. If the functions $\{h_1, h_2, \ldots, h_n\}$ and also the functions $\{k_1, k_2, \ldots, k_n\}$ are linearly independent, then they all have a similar form.*

REFERENCES 11.4.

For results concerning special cases of (11.1) and for further references see e.g. [ACZ3], [ACZ6], [ACZ7], [ACZ10], [ACZ12], [ACZ13], [COI], [CAU], [FLE], [HOS3], [KEM1], [KNP], [KUR3], [OCO1], [PEN], [ROS], [RUK1], [RUK2], [SAT], [SEG], [SZÉ6], [WIL3], [WIL4].

12. Addition and subtraction theorems

The results of Section 10 and Section 11 can be used to study addition and subtraction theorems of general type. Here we consider the problem of addition and subtraction theorems on abelian groups, although corresponding to the previous results in some cases we could consider the more general setting of commutative semigroups. For the sake of simplicity we shall deal with discrete abelian groups only, while the formulation of the corresponding theorems in the nondiscrete case is left to the reader.

Let G be an abelian group, $f : G \to \mathbf{C}$ a complex valued function. By an *addition theorem* for the function f we mean a law, governing how to compute the value of f at $x + y$ if we know its value at x and y. The simplest addition theorems are expressed by the fundamental equations of the additive and of the exponential functions, that is, by the equations

$$f(x + y) = f(x) + f(y),$$

and

$$f(x + y) = f(x)f(y).$$

More generally, we may consider polynomial addition theorems, that is relations, which show how to compute the value of the function f at $x + y$, if we know the values of some fixed functions $g_1, g_2, \ldots, g_n$ at x, and the values of some fixed functions $h_1, h_2, \ldots, h_n$ at y, while for the computation of $f(x + y)$ we use only "polynomial" operations, that is, addition and multiplication. In other words, we call the function $f : G \to \mathbf{C}$ *polynomially additive* (or we say, that f admits a *polynomial addition theorem*), if there exist positive integers n, m, there exists a complex polynomial $P : \mathbf{C}^{n+m} \to \mathbf{C}$ in $n + m$ variables and there are functions $g_i, h_j : G \to \mathbf{C}$ ($i = 1, 2, \ldots, n; j = 1, 2, \ldots, m$) such that

$$(12.1) \qquad f(x + y) = P(g_1(x), g_2(x), \ldots, g_n(x), h_1(y), h_2(y), \ldots, h_m(y))$$

holds for all x, y in G. Obviously, real additive functions and complex exponentials are polynomially additive. More generally, any polynomial of real

101

additive functions and of complex exponentials is polynomially additive. To be more precise, if k, l are positive integers, $Q : \mathbf{C}^{k+l} \to \mathbf{C}$ is a complex polynomial in $k + l$ variables, $a : G \to \mathbf{R}$ are real additive functions, $m_j : G \to \mathbf{C}$ are complex exponentials $(i = 1, 2, \ldots, k; j = 1, 2, \ldots, l)$, and

$$f(x) = Q(a_1(x), a_2(x), \ldots, a_k(x), m_1(x), m_2(x), \ldots, m_l(x))$$

holds for all x in G, then f is polynomially additive. By Theorem 10.1 the converse is also true, that is, a complex valued function on an abelian group is polynomially additive if and only if it is a polynomial of additive functions and exponentials.

THEOREM 12.1. *A complex valued function on an abelian group is polynomially additive if and only if it is a normal exponential polynomial.*

PROOF: First suppose, that f is a normal exponential polynomial. This means, that f has a representation in the form

$$f(x) = \sum_{j=1}^{N} P_j(a_1(x), a_2(x), \ldots, a_k(x))m_j(x)$$

for all x in G, where n is a positive integer, $P_j : \mathbf{R}^k \to \mathbf{C}$ is a complex polynomial in k variables, $m_j : G \to \mathbf{C}$ is a complex exponential $(j = 1, 2, \ldots, N)$ and $a_i : G \to \mathbf{R}$ is additive for $i = 1, 2, \ldots, k$. By the Taylor-formula we have for all x, y in G

$$f(x + y) =$$

$$= \sum_{j=1} \sum_{\alpha_i} \frac{1}{\alpha_1! \ldots \alpha_k!} \partial_1^{\alpha_1} \ldots \partial_k^{\alpha_k} P_j(a_1(x), \ldots, a_k(x))a_1(y)^{\alpha_1} \ldots a_k(y)^{\alpha_k} \times$$

$$\times m_j(x)m_j(y),$$

which shows, that f satisfies a functional equation of the form (10.1) for all x, y in G, with some functions $g_i, h_i : G \to \mathbf{C}$ $(i = 1, 2, \ldots, n)$. This means, that f is polynomially additive.

Conversely, it is obvious, that polynomial additivity of f means, that f satisfies an equation of the form (10.1), which gives by the results of Section 10 that f is a normal exponential polynomial.

As a subtraction theorem for f means a functional equation of the form (12.1) with $f(x - y)$ instead of $f(x + y)$ on the left hand side, which can be

102

reduced to (12.1) by the substitution $-y$ for y, functions possessing a polynomial subtraction theorem are just the normal exponential polynomials too. Using the general method described in Section 10 here we give the general solutions of some classical functional equations expressing addition and subtraction theorems for trigonometric polynomials. These are the functional equations

$$(12.2) \qquad f(x+y) = f(x)g(y) + g(x)f(y),$$

$$(12.3) \qquad f(x+y) = f(x)f(y) - g(x)g(y),$$

$$(12.4) \qquad f(x-y) = f(x)g(y) - g(x)f(y),$$

$$(12.5) \qquad f(x-y) = f(x)f(y) + g(x)g(y)$$

(see [ACZ7]). Substituting $-y$ for y in (12.4) and (12.5) we get

$$(12.4a) \qquad f(x+y) = f(x)\breve{g}(y) - g(x)\breve{f}(y),$$

$$(12.5a) \qquad f(x+y) = f(x)\breve{f}(y) + g(x)\breve{g}(y),$$

and now (12.2), (12.3), (12.4a), (12.5a) are all special cases of (10.7). For the sake of completeness we give here the general solutions of (12.2), (12.3), (12.4), (12.5). The following four theorems are direct consequences of Theorem 10.4.

THEOREM 12.2. *Let G be an abelian group. The functions $f, g : G \to \mathbf{C}$ satisfy the functional equation (12.2) for all x, y in G if and only if they have any of the following forms :*
   (i) $f(x) = 0$, $\qquad g(x) = arbitrary$;
   (ii) $f(x) = a(x)m(x)$, $\qquad g(x) = m(x)$;
   (iii) $f(x) = \alpha(m_1(x) - m_2(x))$, $\qquad g(x) = \frac{1}{2}(m_1(x) + m_2(x))$;
where $m, m_1, m_2 : G \to \mathbf{C}$ are exponentials, $a : G \to \mathbf{C}$ is additive and α is a complex constant.

THEOREM 12.3. *Let G be an abelian group. The functions $f, g : G \to \mathbf{C}$ satisfy the functional equation (12.3) for all x, y in G if and only if they have any of the following forms :*
   (i) $f(x) = (\pm ia(x) + 1)m(x)$, $\qquad g(x) = a(x)m(x)$;

103

(ii) $f(x) = \alpha_1 m_1(x) + \alpha_2 m_2(x), \qquad g(x) = \beta_1 m_1(x) + \beta_2 m_2(x);$
where $m, m_1, m_2 : G \to \mathbf{C}$ are exponentials, $a : G \to \mathbf{C}$ is additive and $\alpha_1, \alpha_2, \beta_1, \beta_2$ are complex constants satisfying

$$\alpha_1^2 - \beta_1^2 = \alpha_1, \qquad \alpha_1 \alpha_2 - \beta_1 \beta_2 = 0, \qquad \alpha_2^2 - \beta_2^2 = \alpha_2.$$

THEOREM 12.4. Let G be an abelian group. The functions $f, g : G \to \mathbf{C}$ satisfy the functional equation (12.4) for all x, y in G if and only if they have any of the following forms :
  (i) $f(x) = 0, \qquad g(x) = arbitrary;$
  (ii) $f(x) = a(x) m_1(x), \qquad g(x) = (\alpha a(x) + 1) m_1(x);$
  (iii) $f(x) = \alpha(m_2(x) - \breve{m}_2(x)), \qquad g(x) = \beta m_2(x) + (1 - \beta) \breve{m}_2(x);$
where $m_1, m_2 : G \to \mathbf{C}$ are exponentials with $m_1^2 = 1$, $a : G \to \mathbf{C}$ is additive and α, β are complex constants.

THEOREM 12.5. Let G be an abelian group. The functions $f, g : G \to \mathbf{C}$ satisfy the functional equation (12.5) for all x, y in G if and only if they have any of the following forms :
  (i) $f(x) = \alpha m_1(x), \qquad g(x) = \beta m_1(x);$
  (ii) $f(x) = \frac{1}{2}(m_2(x) + \breve{m}_2(x)), \qquad g(x) = \pm \frac{i}{2}(m_2(x) - \breve{m}_2(x));$
where $m_1, m_2 : G \to \mathbf{C}$ are exponentials with $m_1^2 = 1$ and α, β are complex constants satisfying

$$\alpha^2 + \beta^2 = \alpha.$$

Our aim now is to apply the results of Section 11. for addition-subtraction theorems.

Let G be an abelian group, $f : G \to \mathbf{C}$ a complex valued function. By a polynomial addition-subtraction theorem for the function f we mean an equation of the form (11.1a) or (11.1b), that is one of the form

$$f(x + y) \pm f(x - y) = \sum_{i=1}^{n} g_i(x) h_i(y),$$

where n is a positive integer, $g_1, g_2, \ldots, g_n$ and $h_1, h_2, \ldots h_n$ are given complex valued functions. Functions having a polynomial addition-subtraction theorem have been characterized in Section 11. Now we give the explicit form of those functions which have a special type of polynomial addition-subtraction theorem. We shall consider the following functional equations:

$$(12.7) \qquad f(x+y) + f(x-y) = g(x),$$

$$(12.8) \qquad f(x+y) + f(x-y) = 2f(x)f(y),$$

$$(12.9) \qquad f(x+y) + f(x-y) = 2f(x)g(y),$$

$$(12.10) \qquad f(x+y) + f(x-y) = g(x)h(y),$$

$$(12.11) \qquad f(x+y) + f(x-y) = 2f(x) + 2f(y),$$

which are well-known special cases of (11.1). Equation (12.7) is a "pexiderization" of the Jensen equation (9.5), equations (12.9) and (12.10) are the generalizations of the d'Alembert equation (12.8) due to Wilson, and equation (12.11) is the square-norm equation considered in Section 9 as (9.6) (see [ACZ7]).

THEOREM 12.6. *Let G be an abelian group. The functions $f, g : G \to \mathbf{C}$ satisfy the functional equation (12.7) for all x, y in G if and only if they have the form :*

$$f(x) = a(x) + \alpha, \qquad g(x) = 2a(x) + 2\alpha,$$

where $a : G \to \mathbf{C}$ is additive and α is a complex constant.

PROOF: For the proof we don't need the previous results, it is elementary. Namely, (12.7) implies with $y = 0$ that $g = 2f$, and with $x = 0$ that $f(x) + f(-x) = 2f(0)$. Then by interchanging x and y in (12.7) and by adding the new equation to (12.7) one gets that $f - f(0)$ is additive. The sufficiency is trivial.

THEOREM 12.7. *Let G be an abelian group. The function $f : G \to \mathbf{C}$ satisfies the functional equation (12.8) for all x, y in G if and only if it has the form*

$$f(x) = \frac{1}{2}(m(x) + m(-x)),$$

105

where $m : G \to \mathbf{C}$ is an exponential.

PROOF: If f is any solution of (12.8), then any φ in $\tau(f)$ satisfies

$$\varphi(x + y) + \varphi(x - y) = 2\varphi(x)f(y)$$

for all x, y in G. If $f \neq 0$, then by spetral synthesis there exists a nonzero complex exponential m in $\tau(f)$, hence we have

$$m(x)(m(y) + m(-y)) = 2m(x)f(y),$$

which implies the necessity. The sufficiency is trivial.

THEOREM 12.8. *Let G be an abelian group. The functions $f, g : G \to \mathbf{C}$ satisfy the functional equation (12.9) for all x, y in G if and only if they have any of the following forms :*
  (i) $f(x) = 0$, $g(x) = arbitrary$;
  (ii) $f(x) = (a(x) + \alpha)m_1(x)$, $g(x) = m_1(x)$;
  (iii) $f(x) = \alpha m_2(x) + \beta \breve{m}_2(x)$, $g(x) = \frac{1}{2}(m_2(x) + \breve{m}_2(x))$;
where $m_1, m_2 : G \to \mathbf{C}$ are exponentials with $m_1^2 = 1$, $a : G \to \mathbf{C}$ is additive and α, β are complex constants.

PROOF: If f is nonzero, then $\tau(f)$ contains a nonzero exponential m, and similarly as above we get

$$g(x) = \frac{1}{2}(m(x) + \breve{m}(x)).$$

If $m = m_1$, where $m_1^2 = 1$, then $sp(f) = m_1$, that is, any exponential monomial in $\tau(f)$ has the form pm_1 with some complex monomial p. Substitution into (12.9) gives
$$p(x + y) + p(x - y) = 2p(x),$$

which is of the form (12.7), and (ii) follows from Theorem 12.6.

If $m = m_2$, where $m_2^2 \neq 1$, then $sp(f) = \{m_2, \breve{m}_2\}$, that is, any exponential monomial in $\tau(f)$ has the form pm_2 or $p\breve{m}_2$, and substitution into (12.9) shows that p is constant. Now (iii) follows from Theorem 4.6.

THEOREM 12.9. *Let G be an abelian group. The functions $f, g : G \to \mathbf{C}$ satisfy the functional equation (12.10) for all x, y in G if and only if they have any of the following forms :*
  (i) $f(x) = 0$, $g(x) = arbitrary$, $h(x) = 0$;

(ii) $f(x) = 0$, $g(x) = 0$, $h(x) = arbitrary$;

(iii) $f(x) = \frac{\gamma}{2}(\alpha a(x) + \beta)m_1(x)$, $g(x) = (\alpha a(x) + \beta)m_1(x)$, $h(x) = \gamma m_1(x)$;

(iv) $f(x) = \alpha\gamma m_2(x) + \beta\gamma\breve{m}_2(x)$, $g(x) = \alpha m_2(x) + \beta\breve{m}_2(x)$, $h(x) = \gamma(m_2(x) + \breve{m}_2(x))$;

where $m_1, m_2 : G \to \mathbf{C}$ are exponentials with $m_1^2 = 1$, $a : G \to \mathbf{C}$ is additive and α, β, γ are complex constants.

PROOF: The proof is very similar to that of the above theorem and is left to the reader.

For equation (12.11) see Theorem 9.7.

Finally we consider one more equation of addition-subtraction type:

$$(12.12) \qquad f(x+y) - f(x-y) = 2f(x)g(y),$$

which also relates to the trigonometric functions.

THEOREM 12.10. Let G be an abelian group. The functions $f, g : G \to \mathbf{C}$ satisfy the functional equation (12.12) for all x, y in G if and only if they have any of the following forms :

(i) $f(x) = 0$, $\qquad g(x) = arbitrary$;

(ii) $f(x) = F(\Phi(x))$, $\qquad g(x) = 0$;

(iii) $f(x) = \alpha m(x)$, $\qquad g(x) = \frac{1}{2}(m(x) - \breve{m}(x))$;

where $m : G \to \mathbf{C}$ is an exponential, $F : G/2G \to \mathbf{C}$ is an arbitrary function, $\Phi : G \to G/2G$ is the natural homomorphism and α is a complex constant.

PROOF: The first two cases are trivial, hence we may suppose that $f \neq 0$ and $g \neq 0$. Then, similarly as in Theorem 12.8, $\tau(f)$ contains a nonzero exponential m and we have

$$g(x) = \frac{1}{2}(m(x) - \breve{m}(x)).$$

It follows, that $m \neq \breve{m}$, and $sp(f) = \{m, \breve{m}\}$. Then any exponential monomial in $\tau(f)$ has the form pm or $q\breve{m}$, and substitution into (12.9) shows that p is constant and q is zero. Now (iii) follows from Theorem 4.6.

REFERENCES 12.11.

For further results and references concerning addition and subtraction theorems see e.g. [ACZ3], [ACZ4], [ACZ6], [ACZ7], [ACZ13], [ACZ14], [ACZ15], [ACZ16], [ACZ18], [COI], [FLE], [GHE1], [GHE3], [HOS1], [HOS3], [ILS1], [ILS2], [ILS3], [KCZ1], [KCZ2], [KCZ3], [KNP], [KUR3], [OCO1], [OCO2], [OCO3], [PEN], [PER], [ROS], [ROT], [RUK1], [RUK2], [SCH], [SEG], [SWI1], [SWI2], [SZÉ6], [SZÉ7], [SZÉ19], [SZÉ24], [UNG1], [UNG2], [VIE], [VIN1], [VIN2], [VIN3], [WIL3], [WIL4].

13. Difference equations on semigroups

In this section functional equations of the form

$$(13.1) \qquad f(x + ny) + \sum_{k=0}^{n-1} c_k(y) f(x + ky) = 0$$

are investigated on commutative semigroups. We note that (13.1) is a common generalization of several well-known functional equations, like the Cauchy equations, the Pexider equation (9.4), the Jensen equation (9.5), equation (9.2), the d'Alembert equation, etc. Concerning these and similar equations see the references listed at the end of this section. If G is a commutative topological semigroup, n a positive integer and $c_k : G \to \mathbf{C}$ are functions ($k = 0, 1, \ldots, n-1$), then the set of all continuous functions $f : G \to \mathbf{C}$ satisfying (13.1) is called the *solution space* of (13.1). Obviously, the solution space of (13.1) is a closed translation invariant linear space, that is, a variety in $\mathcal{C}(G)$. Our aim is to show, that under general conditions on G this variety consists of exponential polynomials. This is not the case in general, and even if it is, the exponential polynomials contained in the solution space are not necessarily normal, as it is shown by the case of equation (9.2).

First we prove the following simple lemma.

LEMMA 13.1. *Let G, H be commutative topological semigroups with identity and with the property, that the solution space of any functional equation of the form (13.1) on G and on H is of finite dimension. Then $G \times H$ has the same property.*

PROOF: Consider the functional equation

$$f(x + nu, y + nv) + \sum_{k=0}^{n-1} c_k(u, v) f(x + ku, y + kv) = 0$$

on $G \times H$, where $f, c_k : G \times H \to \mathbf{C}$ are functions ($k = 0, 1, \ldots n-1$), and f is continuous. Let $v = 0$ in (13.2), then we have that for all y in H the function

108

$x \mapsto f(x, y)$ belongs to the solution space of an equation of the same form on G. Let $\{\Phi_1, \Phi_2, \ldots, \Phi_N\}$ be a basis of this space, then

$$f(x, y) = \sum_{i=1}^{N} \alpha_i(y) \Phi_i(x)$$

holds for all (x, y) in $G \times H$. By the linear independence of the Φ's it follows that there exist elements x_j in G ($j = 1, 2, \ldots, N$) such that the matrix $(\Phi_i(x_j))$ is regular. We have for $j = 1, 2, \ldots, N$

$$f(x_j, y) = \sum_{i=1}^{N} \alpha_i(y) \Phi_i(x_j),$$

which is a system of linear equations for the unknowns $\alpha_i(y)$ ($i = 1, 2, \ldots, N$). The matrix of the system is regular, hence, by Cramer's rule $\alpha_i(y)$ can be expressed as a linear combination of the functions $f(x_1, y), f(x_2, y), \ldots, f(x_N, y)$. Hence α_i is continuous for $i = 1, 2, \ldots, N$. Substitution into (13.2) with $u = 0$ yields

$$\sum_{i=1}^{N} \Phi_i(x) \Big[\alpha_i(y + nv) + \sum_{k=0}^{n-1} c_k(0, v) \alpha_i(y + kv) \Big] = 0$$

for all (x, y) in $G \times H$. By the linear independence of the Φ's it follows, that the functions α_i ($i = 1, 2, \ldots, N$) belong to the solution space of an equation of the form (13.1) on H. If $\{\Psi_1, \Psi_2, \ldots, \Psi_M\}$ is a basis of this space, then we have

$$\alpha_i(y) = \sum_{j=1}^{M} \lambda_{i,j} \Psi_j(y)$$

for all y in H with some constants $\lambda_{i,j}$ in $\mathbf{C}$ ($i = 1, 2, \ldots, N$, $j = 1, 2, \ldots, M$). Now we see, that the functions $(x, y) \mapsto \Phi_i(x) \Psi_j(y)$ for $i = 1, 2, \ldots, N$, $j = 1, 2, \ldots, M$ generate the solution space of (13.2) and the lemma is proved.

On the commutative semigroup $\mathbf{N}$ any nontrivial difference equation of the above type can be written in the form

(13.3)
$$f(x + n) + \sum_{i=0}^{n-1} c_i f(x + i) = 0,$$

where n is a positive integer and c_i is a complex number for $i = 1, 2, \ldots, n$. The polynomial

$$P(\lambda) = \lambda^n + \sum_{i=0}^{n-1} c_i \lambda^i$$

is called the *characteristic polynomial* of (13.3), and its zeros are the characteristic values of (13.3). The following lemma is standard.

LEMMA 13.2. *The function $f : \mathbf{N} \to \mathbf{C}$ is a solution of (13.3) if and only if it is a linear combination of the functions*

$$(13.4) \qquad \varphi_{j,k}(x) = x^j \lambda_k^x \qquad (k = 1, 2, \ldots, m; j = 0, 1, \ldots, n_j - 1),$$

where $\lambda_1, \lambda_2, \ldots, \lambda_m$ are all the different characteristic values of (13.3) with multiplicities $n_1, n_2, \ldots, n_m$, respectively.

PROOF: It is an easy calculation to show, that the functions (13.4) are all solutions of (13.3) on $\mathbf{N}$ (and even on $\mathbf{Z}$), hence any linear combination of them is a solution too. As their number is n, and they are linearly independent, it is enough to show, that the solution space of (13.3) is n dimensional. Obviously, any solution f of (13.3) is uniquely determined by its values $f(0)$, $f(1)$, $\ldots$,$f(n-1)$, and these values can be prescribed arbitrarily. Let f_j denote the unique solution of (13.3) with

$$f_j(k) = \delta_{j,k} \qquad (j, k = 0, 1, \ldots, n-1),$$

then the functions f_j are linearly independent and for any solution f of (13.3) we have

$$f = \sum_{j=0}^{n-1} f(j) f_j,$$

which proves our statement.

LEMMA 13.3. *The function $f : \mathbf{Z} \to \mathbf{C}$ is a solution of (13.3) if and only if it is a linear combination of the functions (13.4).*

PROOF: We have seen in the previous lemma, that the functions (13.4) are all solutions of (13.3) on $\mathbf{Z}$, hence any linear combination of them is a solution too. Let $f : \mathbf{Z} \to \mathbf{C}$ be any solution on $\mathbf{Z}$, then we have

$$f(x) = \sum_{j,k} a_{j,k} \varphi_{j,k}(x)$$

for all x in $\mathbf{N}$, with some complex constants $a_{j,k}$. We show, that this equation holds for any x in $\mathbf{Z}$. Let z be any negative integer, then $\tau_z f$ is a solution of (13.3) on $\mathbf{Z}$, and the functions $\tau_z \varphi_{j,k}$ form a basis of the solution space of (13.3) on $\mathbf{N}$, hence we have for all x in $\mathbf{N}$

$$\tau_z f(x) = \sum_{j,k} a_{j,k} \tau_z \varphi_{j,k}(x),$$

which gives

$$f(z) = \sum_{j,k} a_{j,k} \varphi_{j,k}(z)$$

with $x = 0$.

THEOREM 13.4. *For any positive integer k the solution space of (13.1) on $G = \mathbf{Z}^k$ consists of exponential polynomials.*

PROOF: It is a consequence of Lemma 13.3 that the solution space of (13.1) is finite dimensional in the case $G = \mathbf{Z}$. Hence, by 13.1 the solution space is finite dimensional also in the case $G = \mathbf{Z}^k$. By Theorem 10.1 our statement follows.

We can go one step further by using the following result:

THEOREM 13.5. *Let G, H be discrete abelian groups, $\Phi : G \to H$ an epimorphism, $f : H \to \mathbf{C}$ a function. If $f \circ \Phi$ is an exponential polynomial, then so is f.*

PROOF: Let $F = f \circ \Phi$, then we have

$$F(u) = \sum_{k=1}^{m} P_k(u) M_k(u)$$

for all u in G, with different exponentials M_k, and with arbitrary polynomials P_k $(k = 1, 2, \ldots, m)$. First we prove, that if a function F of the above form is constant on the cosets of $Ker(\Phi)$, then P_k and M_k are constant on the cosets of $Ker(\Phi)$, whenever $P_k \neq 0$ $(k = 1, 2, \ldots, m)$. We prove this by induction on m. In the case $m = 1$ we have

$$F(u) = P_1(u) M_1(u),$$

111

where $P_1 \neq 0$. If P_1 is constant, then the statement is trivial. Suppose, that we have proved it for polynomials with degree at most d, and let $\deg P_1 = d + 1$. Then there exists a w in G for which $\Delta_w P_1 \neq 0$ and we have

$$F(u + w) - M_1(w)F(u) = M_1(w)\Delta_w P_1(u)M_1(u)$$

for all u in G. It is easy to see that the function $u \mapsto F(u + w) - M_1(w)F(u)$ is constant on the cosets of $Ker(\Phi)$, hence, as $M_1(w)\Delta_w P_1(u)$ is a nonzero polynomial of degree at most d, we have that M_1, and then obviously also P_1, is constant on the cosets of Φ. Now let

$$F(u) = \sum_{k=1}^{m+1} P_k(u)M_k(u)$$

for all u in G, where the exponentials M_k are different, and the polynomials P_k are nonzero $(k = 1, 2, \ldots, n)$. Let $N_k = M_k M_1^{-1}$ for $k = 2, 3, \ldots, m + 1$, then we have

$$M_1(u)F(u) = P_1(u) + \sum_{k=2}^{m+1} P_k(u)N_k(u).$$

Obviously, the functions N_k are different exponentials, and they are different from 1. Suppose, that $\deg P_k = d_k$ and P_k has the form

$$P_k(u) = A_k^*(u) + p_k(u) \qquad (k = 1, 2, \ldots, m + 1)$$

for all u in G, where $A_k^* \neq 0$ is a homogeneous polynomial of degree d_k, and p_k is a polynomial of degree at most $d_k - 1$. We show that M_2 is constant on the cosets of $Ker(\Phi)$. Let w be in G such that $M_2(w) \neq M_1(w)$. Now we apply the difference operator Δ_w on both sides of the above equation:

$$\sum_{j=0}^{d_1+1} \binom{d_1 + 1}{j}(M_1(w)^{-1} - 1)\Delta_w^j F[u + (d_1 + 1 - j)w] =$$

$$= \sum_{k=2}^{m+1} M_k(u)[(N_k(w) - 1)^{d_1+1}A_k^*(u) + q_{k,w}(u)]$$

for all u in G, where $q_{k,w}$ is a polynomial of degree at most $d_k - 1$. Here it is easy to check that the left hand side is constant on the cosets of $Ker(\Phi)$. On the other hand, the coefficient of M_2 is nonzero, as $N_2(w) \neq 1$, hence we have, that M_2 and the polynomial $u \mapsto (N_2(w) - 1)^{d_2+1} A_2^*(u) + q_{2,w}(u)$ are constant on the cosets of $Ker(\Phi)$. From this one easily verifies that A_2 is constant on the cosets of $Ker(\Phi)$. By the same argument we get that M_k, A_k are constant on the cosets of $Ker(\Phi)$ for $k = 1, 2, \ldots, m + 1$. Now we can repeat our argument with $F - \sum_{k=1}^{m+1} M_k A_k^*$ instead of F, and continuing this process, we obtain that the polynomials P_k for $k = 1, 2, \ldots, m+1$ are constant on the cosets of $Ker(\Phi)$. Hence, we have that in the representation of F the functions P_k, M_k are constant on the cosets of $Ker(\Phi)$ ($k = 1, 2, \ldots, m$). For any x in H let u be in G such that $\Phi(u) = x$ and we define $p_k(x) = P_k(u)$ and $m_k(x) = M_k(u)$ for $k = 1, 2, \ldots, m$. It is very easy to see that p_k is a polynomial and m_k is an exponential, and hence

$$f(x) = f(\Phi(u)) = \sum_{k=1}^{m} P_k(u) M_k(u) = \sum_{k=1}^{m} p_k(x) m_k(x)$$

is an exponential polynomial.

COROLLARY 13.6. *If G is a finitely generated discrete abelian group, then all solutions f of (13.1) are exponential polynomials.*

PROOF: Let $\Phi : \mathbf{Z}^k \to G$ be an epimorphism, where k is a positive integer. If $F = f \circ \Phi$, then a routine calculation shows that F is a solution of an equation of the form (13.1) on $\mathbf{Z}^k$. Hence, by 13.4, F is an exponential polynomial, and then by 13.5, f is an exponential polynomial.

The following lemma will be useful for the further considerations.

LEMMA 13.7. *If G is a compact connected abelian group, and the different continuous complex nonzero exponentials $m_1, m_2, \ldots, m_n$ of G are given, then there exists an element y in G such that the numbers $m_1(y), m_2(y), \ldots, m_n(y)$ are different.*

PROOF: By the duality theory [HEW], the character group of G is discrete and torsion-free and our statement is equivalent to the following statement: if G is a discrete and torsion-free abelian group and $g_1, g_2, \ldots, g_n$ are different (nonzero) elements of G, then there exists a character χ of G, such that the numbers $\chi(g_k)$ are different for $k = 1, 2, \ldots, n$. As any character of a subgroup of G can be extended to G, we may assume that G is generated by $\{g_1, g_2, \ldots, g_n\}$. Let $\{h_1, h_2, \ldots, h_m\}$ be a maximal independent subset of $\{g_1, g_2, \ldots, g_n\}$ and let $t_1, t_2, \ldots, t_m$ be any numbers in the interval $(0, 1)$ such that $1, t_1, t_2, \ldots, t_m$ are

113

linearly independent over the field of rationals. There exists a character χ of G with the property $\chi(h_k) = 2\pi i t_k$ $(k = 1, 2, \ldots, m)$. On the other hand, if g_k is arbitrary $(k = 1, 2, \ldots, n)$, then there are integers $b \neq 0$ and $a_1, a_2, \ldots, a_m$ such that

$$bg_k = a_1 h_1 + a_2 h_2 + \cdots + a_m h_m.$$

Further, the ratios $a_1/b, a_2/b, \ldots, a_m/b$ are uniquely determined by g_k, and not all a's are zero. Then we let

$$\chi(g_k) = \exp[2\pi i(a_1/b \cdot t_1 + a_2/b \cdot t_2 + \cdots + a_m/b \cdot t_m)].$$

Let $g_l \neq g_k$ and

$$cg_l = d_1 h_1 + d_2 h_2 + \cdots + d_m h_m$$

with some integers $c \neq 0, d_1, d_2, \ldots, d_m$, where not all d's are zero. Hence

$$\chi(g_l) = \exp[2\pi i(d_1/c \cdot t_1 + d_2/c \cdot t_2 + \cdots + d_m/c \cdot t_m)],$$

and $\chi(g_l) = \chi(g_k)$ implies that

$$(ca_1 - bd_1)t_1 + (ca_2 - bd_2)t_2 + \cdots + (ca_m - bd_m)t_m$$

is an integer. By linear independence it follows

$$ca_j = bd_j \qquad (j = 1, 2, \ldots, m),$$

that is

$$bcg_l = cbg_k,$$

which is impossible as $bc \neq 0, g_k \neq g_l$ and G is torsion-free.

THEOREM 13.8. *If $G = \mathbf{R}$, then all continuous solutions f of (13.1) are exponential polynomials.*

PROOF: First we show that at most n different exponentials can belong to the solution space of (13.1). Indeed, the exponential m is a solution of (13.1) if and only if

$$m(y)^n + \sum_{k=0}^{n-1} c_k(y)m(y)^k = 0$$

holds for all y in $\mathbf{R}$. If $n + 1$ different exponentials of $\mathbf{R}$ are given, then there exists an element y in $\mathbf{R}$, at which these exponentials take different values. But this contradicts to the fact, that a complex polynomial of degree n has at most n different zeros. Thus the solution space of (13.1) contains at most n different exponentials, it is not the whole $\mathcal{C}(\mathbf{R})$, hence by definition all continuous solutions of (13.1) are mean periodic functions. Let f be a continuous solution of (13.1) and we consider the Fourier-transform of f as defined in Section 8. Suppose that $\hat{f}(m) \neq 0$ for some exponential m. Then, by Fourier-transformation we obtain

$$\hat{f}(m)(x+ny)m(y)^n + \sum_{k=0}^{n-1} c_k(y)\hat{f}(m)(x+ky)m(y)^k = 0$$

for all x, y in $\mathbf{R}$. The left hand side is a nonzero polynomial in x for all fixed y in $\mathbf{R}$, hence for the leading terms we get

$$m(y)^n + \sum_{k=0}^{n-1} c_k(y)m(y)^k = 0,$$

that is, the exponential m belongs to the solution space of (13.1). By the above considerations we infer that the support of the Fourier-transform of f is finite, and hence f is an exponential polynomial by Corollary 8.8.

THEOREM 13.9. *For any positive integer m, if $G = \mathbf{R}^m$, then all continuous solutions of (13.1) are exponential polynomials.*

PROOF: By Theorem 13.8, the solution space of (13.1) in the case $G = \mathbf{R}$ is finite dimensional. Then by Lemma 13.1, the solution space of (13.1) is finite dimensional also in the case $G = \mathbf{R}^m$, and hence, by Theorem 10.1 it follows that all continuous solutions f of (13.1) are exponential polynomials.

THEOREM 13.10. *If G is a compact, connected abelian group, then all continuous solutions f of (13.1) are trigonometric polynomials.*

PROOF: As G is compact, all continuous solutions f of (13.1) are almost periodic functions. Let $\hat{f}$ denote the Fourier transform of the almost periodic function f, then from (13.1) we obtain

$$\hat{f}(\gamma)\gamma(y)^n + \sum_{k=0}^{n-1} c_k(y)\hat{f}(\gamma)\gamma(y)^k = 0,$$

115

for all y in G and character γ. The polynomial

$$\gamma(y)^n + \sum_{k=0}^{n-1} c_k(y)\gamma(y)^k$$

has at most n roots, and since for any finite set of characters they all differ on a common y by Lemma 13.7, it follows that $\hat{f}$ has finite support, and f is a trigonometric polynomial.

THEOREM 13.11. *Let G be a locally compact compactly generated abelian group in which the set of all compact elements is connected. Then all continuous solutions f of (13.1) are exponential polynomials.*

PROOF: By the structure theory of locally compact compactly generated abelian groups, G has the form ([HEW], §9, Theorem 9.8)

$$G = \mathbf{R}^m \oplus \mathbf{Z}^k \oplus K$$

where K is a compact abelian group, which is now connected, by our assumption. Our statement follows from 13.1, 13.4, 13.9 and 13.10.

COROLLARY 13.12. *Let G be as in the previous theorem. Then a continuous complex valued function f on G is an exponential polynomial if and only if f is a solution of a functional equation of the form (13.1). In this case the functions c_k ($k = 0, 1, \ldots, n-1$) can be chosen to be continuous.*

PROOF: The sufficiency of the condition follows from 13.11. The necessity follows from Theorem 8.2 if in the proof we substitute $y_1 = y_2 = \cdots = y_N = y$.

By using the Fourier-transform of exponential polynomials as defined in Section 7, the problem of determining all exponential polynomial solutions of a functional equation of the form (13.1) can be reduced to the problem of determining all polynomial solutions of an equation of the same form. The latter problem can be solved by using Theorem 2.3 and comparing the homogeneous terms of the same degree.

Now we give a simple example for a compact abelian group G such that not all continuous solutions of an equation of the form (13.1) are exponential polynomials. Let G be the direct product of countably many copies of the set $\{0, 1\}$ with addition mod 2. Then with the product topology, G is a compact abelian group. We represent the elements x of G as infinite sequences $x = \{x_k\}$ where $x_k = 0$ or 1 for $k = 1, 2, \ldots$. If χ is any character of G, then by (23.21) in [HEW], there exists a positive integer N such that $\chi(x) = \chi(y)$ for any x, y in G with the property $x_k = y_k$ for $k \leq N$. Then obviously for any trigonometric polynomial on G there exists a positive integer with the same property. On the other hand, it is trivial, that $2y = y + y = 0$ holds for all y in G, 0 being the identity of G. It follows, that all continuous functions f on G are solutions of the functional equation

$$f(x + 2y) = f(x),$$

which is of the form (13.1). But it is easy to see by the above considerations that there are continuous functions on G which are not trigonometric polynomials, hence not all continuous solutions f of this equation are exponential polynomials.

REFERENCES 13.15.

For further results and references concerning this section see [ACZ7], [EDG], [GAJ5], [GHE3], [LEF], [MAU], [SZÉ4], [SZÉ11], [SZÉ16], [SZÉ24].

14. Mean-value type functional equations

In this section we deal with functional equations of mean-value type; that is, with the functional equations

$$(14.1) \qquad \left[\sum_{i=1}^{n}(\tau_t^i + \tau_{-t}^i)\right] f = 2nf,$$

and

$$(14.2) \qquad \left[\prod_{i=1}^{n}(\tau_t^i + \tau_{-t}^i)\right] f = 2^n f,$$

where an abelian group G is given, n is a positive integer, $f : G^n \to \mathbf{C}$ is a function and τ_t^i denotes the translation operator in the i -th variable with increment t, that is,

$$\tau_t^i f(x_1, x_2, \ldots, x_n) = f(x_1, x_2, \ldots, x_{i-1}, x_i + t, x_{i+1}, \ldots, x_n)$$

holds for $i = 1, 2, \ldots, n$ and for all $x_1, x_2, \ldots, x_n, t$ in G. Equation (14.1), resp. (14.2) is called *octahedron*, resp. *cube equation* by obvious geometrical reasons. For $n = 1$ they coincide and they are equivalent to the Jensen equation. Equations of the above and similar type have been dealt with by several authors (see the list of references at the end of this section). Special cases of (14.1) and (14.2) have been solved under and without regularity conditions. Further it has been proved (see [MAC], [SWE]) that (14.1) implies (14.2) for any n, and (14.2) implies (14.1) for $n \leq 4$. It has been conjectured by D.Z.Djokovič and H.Haruki (see [SWE]) that (14.1) and (14.2) are equivalent for all n. Here we prove this conjecture. Another conjecture of H.Haruki (see [HHA1]) is, that any continuous complex valued solution of (14.1) and (14.2) on $\mathbf{R}^n$ is a linear combination of all partial derivatives of a given harmonic polynomial Q_n (see below). Here we prove this conjecture too. We note that our method can be applied also for other functional equations of mean-value type, which have been studied extensively.

118

To deal with the general case, we need the following simple lemmata.

LEMMA 14.1. *Let G be an abelian group. Then any nonzero complex exponential on G is an extremal point of the convex hull of all nonzero complex exponentials on G.*

PROOF: It is obviously enough to prove the following statement: if k is any positive integer, $m_1, m_2, \ldots, m_k$ are nonzero complex exponentials and $\lambda_1, \lambda_2, \ldots, \lambda_k$ are positive real numbers with $\lambda_1 + \lambda_2 + \cdots + \lambda_k = 1$, then

$$\sum_{i=1}^{k} \lambda_i m_i = 1$$

implies $m_1 = m_2 = \cdots = m_k = 1$. This is trivial for $k = 1$. If $k \geq 2$ and $m_1, m_2, \ldots, m_k$ are different, then the statement follows from the fact, that different nonzero exponentials are linearly independent. If $m_1, m_2, \ldots, m_k$ are not different, then the statement follows by induction on k.

LEMMA 14.2. *Let n, l be positive integers and let $a_{i,j}$ be real numbers for $i = 1, 2, \ldots, l; j = 1, 2, \ldots, n$. Then we have*

$$\sum_{\epsilon \in \{-1,1\}^n} \left[\sum_{j_1=1}^{n} \epsilon_{j_1} a_{1,j_1} \right] \left[\sum_{j_2=1}^{n} \epsilon_{j_2} a_{2,j_2} \right] \cdots \left[\sum_{j_l=1}^{n} \epsilon_{j_l} a_{l,j_l} \right] =$$

$$= \begin{cases} 0, & \text{if } l \text{ is odd;} \\ 2^n \sum_{j=1}^{n} a_{1,j} a_{2,j} \ldots a_{l,j}, & \text{if } l \text{ is even.} \end{cases}$$

(Here we used the notation $\epsilon = (\epsilon_1, \epsilon_2, \ldots, \epsilon_n)$.)

PROOF: The above expression can be rewritten in the form

$$\sum_{\epsilon \in \{-1,1\}^n} \left[\sum_{j_1=1}^{n} \epsilon_{j_1} a_{1,j_1} \right] \left[\sum_{j_2=1}^{n} \epsilon_{j_2} a_{2,j_2} \right] \cdots \left[\sum_{j_l=1}^{n} \epsilon_{j_l} a_{l,j_l} \right] =$$

$$\sum_{\epsilon \in \{-1,1\}^n} \sum_{j_1, j_2, \ldots, j_l = 1}^{n} \epsilon_{j_1} \epsilon_{j_2} \cdots \epsilon_{j_l} a_{1,j_1} a_{2,j_2} \ldots a_{l,j_l} =$$

$$\sum_{j_1, j_2, \ldots, j_l = 1}^{n} \left[\sum_{\epsilon \in \{-1,1\}^n} \epsilon_{j_1} \epsilon_{j_2} \cdots \epsilon_{j_l} \right] a_{1,j_1} a_{2,j_2} \ldots a_{l,j_l}.$$

We may obviously suppose, that $n \geq 2$. We consider any $p \neq q$ and a term of the form

$$a_{1,j_1} \ldots a_{p,j_p} \ldots a_{q,j_q} \ldots a_{l,j_l}.$$

If $j_p \neq j_q$, then the sum contains this term with the following 4 coefficients

$$\sum_{\epsilon \in \{-1,1\}^n, \epsilon_{j_p}=1, \epsilon_{j_q}=1} \epsilon_{j_1} \epsilon_{j_2} \ldots \epsilon_{j_l}, \qquad \sum_{\epsilon \in \{-1,1\}^n, \epsilon_{j_p}=-1, \epsilon_{j_q}=1} \epsilon_{j_1} \epsilon_{j_2} \ldots \epsilon_{j_l},$$

$$\sum_{\epsilon \in \{-1,1\}^n, \epsilon_{j_p}=1, \epsilon_{j_q}=-1} \epsilon_{j_1} \epsilon_{j_2} \ldots \epsilon_{j_l}, \qquad \sum_{\epsilon \in \{-1,1\}^n, \epsilon_{j_p}=-1, \epsilon_{j_q}=-1} \epsilon_{j_1} \epsilon_{j_2} \ldots \epsilon_{j_l},$$

the sum of which is obviously zero. Hence the sum contains terms of the above form with nonzero coefficient only for $j_1 = j_2 = \cdots = j_l$, with the coefficient

$$\sum_{\epsilon \in \{-1,1\}^n} \epsilon_{j_1} \epsilon_{j_2} \ldots \epsilon_{j_l} = \sum_{\epsilon \in \{-1,1\}^n} (\epsilon_{j_1})^l,$$

which proves our statement.

THEOREM 14.3. *Let G be an abelian group and let n be a positive integer. The function $f : G^n \to \mathbf{C}$ is a solution of (14.1) if and only if it is a solution of (14.2).*

PROOF: By the results of Section 8 in order to show the equivalence of (14.1) and (14.2) we may suppose that G is finitely generated. Obviously all solutions of (14.1), resp. (14.2) form a variety, hence by spectral synthesis it is enough to show, that a normal exponential monomial $\varphi : G^n \to \mathbf{C}$ is a solution of (14.1) if and only if it is a solution of (14.2). First we show, that any normal exponential monomial solution of (14.1), resp. (14.2) is a monomial, that is, the spectrum of (14.1), resp. (14.2) consists of the only exponential $m = 1$. By Theorem 6.11 any complex exponential m on G^n has the form

$$m(x_1, x_2, \ldots, x_n) = m_1(x_1) m_2(x_2) \ldots m_n(x_n),$$

where m_i is an exponential on G for $i = 1, 2, \ldots, n$. Substituting m into (14.1), resp. (14.2), we have that the exponential $m_0 = 1$ is a convex combination of exponentials, hence we can apply Lemma 14.1, which immediately implies that $m = 1$.

120

By Theorem 6.10 any monomial on G^n is a constant multiple of the function of form

$$\varphi(x_1, x_2, \ldots, x_n) = \prod_{j=1}^{l} \sum_{i=1}^{n} a_{j,i}(x_i)$$

where l is a positive integer and $a_{i,j} : G \to \mathbf{R}$ is additive for $i = 1, 2, \ldots, n$; $j = 1, 2, \ldots, k$. Supposing that the additive functions

$$(x_1, x_2, \ldots, x_n) \mapsto \sum_{i=1}^{n} a_{j,i}(x_i)$$

are linearly independent for different j, we write φ in the form

$$\varphi(x_1, x_2, \ldots, x_n) = \prod_{j=1}^{k} \left[\sum_{i=1}^{n} a_{j,i}(x_i) \right]^{\alpha_j},$$

where $\alpha = (\alpha_1, \alpha_2, \ldots, \alpha_k)$ is an arbitrary multi-index and k is a positive integer. First we show, that if φ is a solution of (14.1) or (14.2), then $\alpha_j \leq 1$ for $j = 1, 2, \ldots, k$. Indeed, if $\alpha_j > 1$ for some j, then by Lemma 2.8 the monomial

$$\psi_j(x_1, x_2, \ldots, x_n) = \left[\sum_{i=1}^{n} a_{j,i}(x_i) \right]^2$$

is a solution of (14.1), resp. (14.2). By substitution into (14.1), resp. (14.2), we obtain with $x_1 = x_2 = \cdots = x_n = 0$, that

$$\sum_{i=1}^{n} a_{j,i}(t)^2 = 0,$$

respectively

$$\sum_{\epsilon \in \{-1,1\}^n} \left[\sum_{i=1}^{n} \epsilon_i a_{j,i}(t) \right]^2 = 0,$$

holds for all t. The first equation trivially implies that $a_{j,i} = 0$ for all j, i, which contradicts to the linear independence. The same holds for the second equation, as

$$\sum_{i=1}^{n} \epsilon_i a_{j,i}(t) = 0$$

follows for all j and for all $\epsilon = (\epsilon_1, \epsilon_2, \ldots, \epsilon_n)$ in $\{-1, 1\}^n$. Hence any nonzero monomial solution of (14.1), resp. (14.2) has the form

$$\varphi(x_1, x_2, \ldots, x_n) = \prod_{j=1}^{k} \left[\sum_{i=1}^{n} a_{j,i}(x_i) \right],$$

where the additive functions $(x_1, x_2, \ldots, x_n) \mapsto \sum_{i=1}^{n} a_{j,i}(x_i)$ are linearly independent for different j. Now we derive necessary and sufficient conditions for φ in order that it is a solution of (14.1). For the sake of brevity we use the notation

$$\xi_j = \sum_{i=1}^{n} a_{j,i}(x_i)$$

for $j = 1, 2, \ldots, k$, supposing that $x_1, x_2, \ldots, x_n, t$ are fixed in G. Substitution into (14.1) gives

$$\sum_{i=1}^{n} \left[\prod_{j=1}^{k} (\xi_j + a_{j,l}(t)) + \prod_{j=1}^{k} (\xi_j - a_{j,l}(t)) \right] = 2n \prod_{j=1}^{k} \xi_j.$$

By using the linear independence of the ξ's and comparing the coefficients of the products of them we get the necessary and sufficient condition for the validity of this equation as follows:

(14.1a)
$$\sum_{l=1}^{n} a_{j_1,l}(t) a_{j_2,l}(t) \ldots a_{j_s,l}(t) = 0$$

for any even $s \leq k$ and for all t in G, whenever $j_1, j_2, \ldots, j_s$ are different. That is, (14.1a) is necessary and sufficient for φ is a solution of (14.1). Substituting φ into (14.2) we get

$$\sum_{\epsilon \in \{-1,1\}^n} \prod_{j=1}^{k} (\xi_j + \sum_{l=1}^{n} \epsilon_l a_{j,l}(t)) = 2^n \prod_{j=1}^{k} \xi_j.$$

Here we use the linear independence of the ξ's again and compare the coefficients of their products which are of the form as in Lemma 14.2. We infer that the necessary and sufficient condition for the validity of this equation is (14.1a) again, and hence our theorem is proved.

122

From now on it is enough to deal with (14.1) only, which has a somewhat simpler structure than (14.2). Now we describe the C^∞ solutions of (14.1) in the case $G = \mathbf{R}$. In order to do this we need some notation.

Let $k, \alpha_1, \alpha_2, \ldots, \alpha_k$ be positive integers, then we let

$$P_{\alpha_1,\alpha_2,\ldots,\alpha_k}^{(k)}(\partial) = \sum_{1 \leq i_1 \neq \cdots \neq i_k \leq n} \partial_{i_1}^{2\alpha_1} \partial_{i_2}^{2\alpha_2} \ldots \partial_{i_k}^{2\alpha_k}.$$

This means, that we have, for instance

$$P_\alpha^{(1)}(\partial) = \sum_{i=1}^n \partial_i^{2\alpha}$$

for all positive integers α, and

$$P_{1,1,\ldots,1}^{(n)} = \partial_1^2 \partial_2^2 \ldots \partial_n^2.$$

LEMMA 14.4. *Let n be a positive integer. Then any C^∞ solution $f : \mathbf{R}^n \to \mathbf{C}$ of (14.1) satisfies*

$$P_{\alpha_1,\alpha_2,\ldots,\alpha_k}^{(k)} f = 0$$

for all positive integers $k, \alpha_1, \alpha_2, \ldots, \alpha_k$.

PROOF: If we differentiate (14.1) 2α times with respect to t, and substitute $t = 0$, then we have

$$P_\alpha^{(1)}(\partial)f = 0,$$

which is our statement for $k = 1$. Now we consider the product

$$(\partial_1^{2\alpha_1} + \cdots + \partial_n^{2\alpha_1}) \ldots (\partial_1^{2\alpha_k} + \cdots + \partial_n^{2\alpha_k})(\partial_1^{2\alpha_{k+1}} + \cdots + \partial_n^{2\alpha_{k+1}}).$$

Expanding it we see that it is a sum of terms of the form

$$P_{\beta_1,\beta_2,\ldots,\beta_i}^{(i)}(\partial),$$

where $1 \leq i \leq k + 1$, the β's are sums of some α's with

$$\beta_1 + \beta_2 + \cdots + \beta_i = \alpha_1 + \alpha_2 + \cdots + \alpha_{k+1},$$

further the only term with $i = k + 1$ is

$$P_{\alpha_1,\alpha_2,\ldots,\alpha_{k+1}}^{(k+1)}(\partial).$$

On the other hand, the above product applied on f is obviously zero by our statement for $k = 1$, hence it follows by induction

$$P^{(k+1)}_{\alpha_1,\alpha_2,\ldots,\alpha_{k+1}}(\partial)f = 0,$$

and the statement is proved.

LEMMA 14.5. *Let n be a positive integer. Then any C^∞ solution $f : \mathbf{R}^n \to \mathbf{C}$ of (14.1) satisfies*

$$\partial_i^{2n} f = 0$$

for any $i = 1, 2, \ldots, n$.

PROOF: First we prove by induction on k, that

$$\partial_1^{2k}\left(\sum_{1 < i_1 < \cdots < i_{n-k} \leq n} \partial_{i_1}^2 \partial_{i_2}^2 \ldots \partial_{i_{n-k}}^2 \right)f = 0$$

holds for $k = 1, 2, \ldots, n$. For $k = 1$ this follows from Lemma 14.4, as

$$P^{(n)}_{1,1,\ldots,1}(\partial)f = 0.$$

On the other hand, Lemma 14.4 also implies

$$\sum_{1 \leq i_1 < \cdots < i_{n-k} \leq n} \partial_{i_1}^2 \partial_{i_2}^2 \ldots \partial_{i_{n-k}}^2 f = 0,$$

that is,

$$\partial_1^2 \left(\sum_{1 < i_1 < \cdots < i_{n-k-1} \leq n} \partial_{i_1}^2 \partial_{i_2}^2 \ldots \partial_{i_{n-k-1}}^2 \right)f +$$

$$+ \sum_{1 < i_1 < \cdots < i_{n-k} \leq n} \partial_{i_1}^2 \partial_{i_2}^2 \ldots \partial_{i_{n-k}}^2 f = 0.$$

Applying ∂_1^{2k} on both sides we get our statement by induction on k. Now we have with $k = n$

$$\partial_1^{2n} f = 0,$$

which is the statement of the lemma for $i = 1$. By symmetry our lemma is proved.

For any positive n we use the notation

$$Q_n(x_1, x_2, \ldots, x_n) = x_1 x_2 \ldots x_n \prod_{i<j}(x_i^2 - x_j^2),$$

whenever $(x_1, x_2, \ldots, x_n)$ is in $\mathbf{R}^n$.

THEOREM 14.6. *Let n be a positive integer. Any C^∞ solution of (14.1) belongs to $\tau(Q_n)$.*

PROOF: It can be checked by direct computation that Q_n is a solution of (14.1), hence any translate of Q_n is a solution, too. On the other hand, by the Taylor-formula, $\tau(Q_n)$ is generated by the partial derivatives of Q_n, hence it is enough to show, that any polynomial solution of (14.1) is a linear combination of partial derivatives of Q_n. We prove this statement by induction on n, and it is trivial for $n = 1$. Supposing that it has been proved for all values not greater than n, we prove it for $n + 1$. By Lemma 14.5 we have

$$\partial_{n+1}^{2n+2} f = 0,$$

and hence $\partial_{n+1}^{2n+1} f$ is independent of x_{n+1}. We have

$$\partial_{n+1}^{2n+1} f(x_1, x_2, \ldots, x_n, x_{n+1}) = \varphi_0(x_1, x_2, \ldots, x_n),$$

and obviously φ_0 is also a solution of (14.1). Substitution gives that φ_0 is actually a solution on $\mathbf{R}^n$, and then by induction we have

$$\varphi_0(x_1, x_2, \ldots, x_n) = \sum_\alpha c_\alpha^{(0)} \partial_1^{\alpha_1} \partial_2^{\alpha_2} \ldots \partial_n^{\alpha_n} Q_n(x_1, x_2, \ldots, x_n)$$

with some constants $c_\alpha^{(0)}$. On the other hand we observe that

$$Q_n(x_1, x_2, \ldots, x_n) = \frac{(-1)^n}{(2n+1)!} \partial_{n+1}^{2n+1} Q_{n+1}(x_1, x_2, \ldots, x_n, x_{n+1})$$

and thus

$$\varphi_0(x_1, x_2, \ldots, x_n) =$$

$$= \sum_\alpha c_\alpha^{(0)} \frac{(-1)^n}{(2n+1)!} \partial_1^{\alpha_1} \ldots \partial_n^{\alpha_n} \partial_{n+1}^{2n+1} Q_{n+1}(x_1, x_2, \ldots, x_n, x_{n+1}).$$

125

Now let

$$S_0(x_1, x_2, \ldots, x_{n+1}) =$$

$$= \sum_\alpha c_\alpha^{(0)} \frac{(-1)^n}{(2n+1)!} \partial_1^{\alpha_1} \ldots \partial_n^{\alpha_n} Q_{n+1}(x_1, x_2, \ldots, x_n, x_{n+1}),$$

then S_0 is a solution of (14.1), further $\partial_{n+1}^{2n+1}(f - S_0) = 0$. If we define $f_1 = f - S_0$, then f_1 is a solution of (14.1) and $\partial_{n+1}^{2n+1} f_1 = 0$. Hence

$$\partial_{n+1} f_1(x_1, x_2, \ldots, x_n, x_{n+1}) = \varphi_1(x_1, x_2, \ldots, x_n),$$

and continuing this process we arrive at $f_{2n+1} = f_{2n} - S_{2n}$, where

$$\partial_{n+1} f_{2n+1} = 0,$$

and $S_0, S_1, \ldots, S_{2n}$ belong to $\tau(Q_{n+1})$, further $f_{k+1} = f_k - S_k$ is a solution of (14.1) for $k = 1, 2, \ldots, 2n$. It follows

$$f_{2n+1}(x_1, x_2, \ldots, x_n, x_{n+1}) = \varphi_{2n+1}(x_1, x_2, \ldots, x_n)$$

and we have again that f_{2n+1} belongs to $\tau(Q_{n+1})$ by induction. Finally we obtain

$$f = S_0 + S_1 + \cdots + S_{2n} + f_{2n+1},$$

and hence our theorem is proved.

THEOREM 14.7. *Let n be a positive integer. A locally integrable function $f : \mathbf{R}^n \to \mathbf{C}$ is a solution of (14.1) or of (14.2) if and only if it is almost everywhere equal to a linear combination of the partial derivatives of Q_n.*

PROOF: The equation (14.1) obviously implies for any locally integrable function $f : \mathbf{R}^n \to \mathbf{C}$ that f satisfies the Laplace differential equation in the distribution sense, hence it is C^∞. Then the statement is a consequence of the previous results.

REFERENCES 14.8.

For further results and references concerning mean-value type functional equations see [ACZ8], [ALR1], [ALR2], [BAK1], [DJO2], [ETI], [GIR2], [HEM], [HAS1], [HAS2], [HAS3], [HAS5], [HHA1], [HHA2], [LAI3], [LAI4], [MAC], [MCK4], [MIL], [SWE].

15. Applications for differential equations

This section is devoted to some special applications of the Fourier-transform of exponential polynomials for ordinary and for partial differential equations. First we describe the behavior of the Fourier-transformation with respect to differentiation. We note, that ∂ denotes the gradient operator $(\partial_1, \partial_2, \ldots, \partial_n)$ and I is the identity operator. If P is a complex polynomial in n variables, then $P(\partial)$ has the usual meaning, as in the theory of differential operators.

THEOREM 15.1. *Let n be a positive integer, P a complex polynomial in n variables and let $f : \mathbf{R}^n \to \mathbf{C}$ be an exponential polynomial. Then we have for any exponential $m : \mathbf{R}^n \to \mathbf{C}$*

$$\big(P(\partial)f\big)\hat{}(m) = P\big(\partial + \partial(0)I\big)\hat{f}(m).$$

PROOF: It is enough to show that

$$\big(\partial^\alpha f\big)\hat{}(m) = \big(\partial + \partial m(0)I\big)^\alpha \hat{f}(m)$$

holds for each multi-index α. By induction on $|\alpha|$ it is enough to prove that for $j = 1, 2, \ldots, n$ we have

$$\big(\partial_j f\big)\hat{}(m) = \big(\partial_j + \partial_j m(0)I\big)\hat{f}(m).$$

But this follows by the properties of the invariant mean M (see Section 7) from the equation

$$\big(\partial_j f\big)\hat{}(m) = M\big(\partial_j f\check{m}\big) = M\big[\partial_j(f\check{m}) - f\partial_j\check{m}\big] =$$

$$= M\big[\partial_j(f\check{m})\big] - M\big(f\partial_j\check{m}\big) = \partial_j M(f\check{m}) - M\big[f\big(-\partial_j m(0)\big)\check{m}\big] =$$

127

$$= \partial_j \hat{f}(m) + \partial_j m(0) \hat{f}(m) = \left(\partial_j + \partial_j m(0)I \right) \hat{f}(m)$$

as M clearly commutes with partial differentiation.

In the special case $n = 1$ let D denote the operator of (ordinary) differentiation, then we have

COROLLARY 15.2. *Let* $f : \mathbf{R} \to \mathbf{C}$ *be an exponential polynomial. Then, for any exponential* $m : \mathbf{R} \to \mathbf{C}$ *and for any nonnegative integer* k *we have*

$$\left(D^k f \right)\hat{}(m) = \left(D + m'(0)I \right)^k \hat{f}(m).$$

We note, that in the case $G = \mathbf{R}^n$ the nonzero complex exponentials have the form $m(x) = \exp <\lambda, x>$ with $\lambda \in \mathbf{C}^n$, hence the generalized character group can be identified with $\mathbf{C}^n$. Then the above formula can be rewritten as

$$\left(D^k f \right)(\lambda) = \left(D + \lambda \right)^k \hat{f}(\lambda),$$

agreeing that λ stands for the operator λI. Similarly, the formula in Theorem 15.1 can be written as

$$\left(P(\partial)f \right)\hat{}(\lambda) = P\left(\partial + \lambda \right) \hat{f}(\lambda).$$

APPLICATION 15.3.

We consider the inhomogeneous linear differential equation with constant coefficients

(15.1) $$P(D)y = f,$$

where P is a complex polynomial of degree n and f is an exponential polynomial. It is easy to see that all solutions y of this equation are exponential polynomials and we show how to determine them without integration.

Suppose, that y is an exponential solution of (15.1), then by Fourier-transformation we have

(15.2) $$P(D + \lambda)\hat{y}(\lambda) = \hat{f}(\lambda)$$

for all λ in $\mathbf{C}$. Here $\hat{y}(\lambda)$, $\hat{f}(\lambda)$ are polynomials and $\hat{f}(\lambda) = 0$ with the exception of a finite set. Hence the problem of solving (15.1) has been reduced to the problem of finding all polynomial solutions q of an equation of the form

$$(15.3) \qquad\qquad P(D + \lambda)q = p,$$

where p is a given polynomial. This latter problem is, however, trivial to solve, by comparing the coefficients of $P(D + \lambda)q$ and p. Hence we get

THEOREM 15.4. *Let n be a positive integer, $P : \mathbf{R} \to \mathbf{C}$ a polynomial of degree n and $f : \mathbf{R} \to \mathbf{C}$ an exponential polynomial. Then all solutions y of (15.1) are exponential polynomials. Further, y is a solution if and only if for any λ in $\mathbf{C}$, which is a zero of order k of P, the coefficients b_j of the polynomial $D^k \hat{y}(\lambda)$ satisfy the system*

$$(15.4) \qquad\qquad c_i = \sum_{j=0}^{min(n-k, N-i)} \frac{P^{(k+j)}(\lambda)}{(k+j)!} j! \binom{i+j}{j} b_{i+j}$$

for $i = 0, 1, \ldots, N$, where the c_i's are the coefficients and N is the degree of $\hat{f}(\lambda)$.

In the case $\hat{f}(\lambda) = 0$ it follows from (15.4) that $\hat{y}(\lambda) \neq 0$ implies $P(\lambda) = 0$, hence λ is a characteristic value of (15.1) with multiplicity $k \geq 1$. Then by (15.4) it follows $D^k \hat{y}(\lambda) = 0$ and hence $\hat{y}(\lambda)$ is an arbitrary polynomial of degree at most $k - 1$. By Theorem 15.4 we can solve (15.1) as follows. First we determine the zeros of P with their multiplicities. Suppose, that the support of f is the finite set $\lambda_1, \lambda_2, \ldots, \lambda_l$. Then we determine those value of k for which $P^{(j)}(\lambda_i) = 0$ $(j = 0, 1, \ldots, k - 1)$ and $P^{(k)}(\lambda_i) \neq 0$, and we solve the system of equations (15.4) with $\lambda = \lambda_i$. The constants c_j $(j = 0, 1, \ldots, N)$ are the coefficients of $\hat{f}(\lambda_i)$. From the solution we have the polynomial $q_i = \hat{y}(\lambda_i)$. Then the general solution of (15.1) is the following

$$y(x) = \sum_{j=1}^{s} p_j(x) e^{\mu_j x} + \sum_{i=1}^{l} q_i(x) e^{\lambda_i x},$$

where $\mu_1, \mu_2, \ldots, \mu_s$ are the zeros of P with multiplicities $n_1, n_2, \ldots, n_s$ $(n_1 + n_2 + \cdots + n_s = n)$, and p_j is an arbitrary polynomial of degree at most $n_j - 1$ $(j = 1, 2, \ldots, s)$.

We note that the same method can be applied to determine all exponential polynomial solutions of inhomogeneous linear differential equations with polynomial coefficients.

EXAMPLE 15.5.

We solve the equation

$$y'' - y = x^2 e^x - x \cos x + 1.$$

The characteristic polynomial is $P(\lambda) = \lambda^2 - 1$, the characteristic values are $\mu_1 = 1, \mu_2 = -1$. By Fourier-transformation we have

$$q'' + 2\lambda q' + (\lambda^2 - 1)q = \begin{cases} x^2 & \textit{if} \quad \lambda = 1, \\ -\frac{x}{2} & \textit{if} \quad \lambda = i, \\ -\frac{x}{2} & \textit{if} \quad \lambda = -i, \\ 1 & \textit{if} \quad \lambda = 0, \\ 0 & \textit{otherwise.} \end{cases}$$

Here we used the notation $q = \hat{y}(\lambda)$. For $\lambda = 1$ we have from (15.4)

$$1 = 2b_2, \quad 0 = 2b_1 + 2b_2, \quad 0 = 2b_0 + b_1,$$

hence

$$\hat{y}(1)(x) = \frac{1}{6}x^3 - \frac{1}{4}x^2 + \frac{1}{4}x + c.$$

Similarly, we have

$$\hat{y}(i)(x) = \frac{1}{4}x + \frac{i}{4},$$

$$\hat{y}(-i)(x) = \frac{1}{4}x - \frac{i}{4},$$

$$\hat{y}(0)(x) = -1.$$

Finally, if $\lambda \neq 1, \lambda \neq \pm i, \lambda \neq 0$, then $\hat{y}(\lambda) \neq 0$ implies $\lambda = -1$ and $\hat{y}(\lambda)(x) = d$, a constant. Thus the general solution of the equation is

$$y(x) = ce^x + de^{-x} + \left(\frac{1}{6}x^3 - \frac{1}{4}x^2 + \frac{1}{4}x\right)e^x + \frac{1}{2}x \cos x - \frac{1}{2}x \sin x - 1.$$

EXAMPLE 15.6.

We determine all exponential polynomial solutions of the equation

$$(x^2 - 1)y'' - (3x + 1)y' - (x^2 - x)y = 0.$$

By Fourier-transformation we have

$$(x^2 - 1)q'' + (2\lambda x^2 - 3x - 2\lambda - 1)q' + (x^2(\lambda^2 - 1) - x(3\lambda - 1) - \lambda^2 - \lambda)q = 0,$$

where $q = \hat{y}(\lambda)$. By comparing the coefficients we have that $q \neq 0$ implies $\lambda^2 = 1$. If $\lambda = 1$, then

$$(x^2 - 1)q'' + (2x^2 - 3x - 3)q' - (2x + 2)q = 0,$$

and it follows, that $q \neq 0$ implies $\deg q = 1$, which leads to a contradiction. Hence $\hat{y}(1) = 0$. If $\lambda = -1$, then

$$(x^2 - 1)q'' + (-2x^2 - 3x + 1)q' + 4xq = 0,$$

and $q \neq 0$ implies $\deg q = 2$. Substitution gives $q(x) = (x + 1)^2$. Hence an exponential polynomial solution of the equation is

$$y(x) = (x + 1)^2 e^{-x}.$$

Using this solution the equation can be reduced to a linear equation of first order by a standard method.

APPLICATION 15.7.

Now we consider the Cauchy-problem for the heat equation

(15.5)
$$\begin{cases} \partial_t u = a^2 \Delta u, \\ u(x, 0) = u_0(x). \end{cases}$$

Here we suppose that $u_0 : \mathbf{R}^n \to \mathbf{C}$ is an exponential polynomial and we try to find an exponential polynomial $u : \mathbf{R}^n \times \mathbf{R} \to \mathbf{C}$ which is a solution of the above problem. It is known that our problem has at most one solution. We note that ∂_t denotes the differential operator ∂_{n+1} on $\mathbf{R}^n \times \mathbf{R}$, and $\Delta = \sum_{i=1}^n \partial_i^2$ is the Laplacian on $\mathbf{R}^n$.

By taking the Fourier-transform of both sides of the first equation in (15.5) we obtain

$$\partial_t \hat{u}(\lambda, \mu) + \mu \hat{u}(\lambda, \mu) = a^2(\Delta + 2 < \lambda, \partial > + < \lambda, \lambda >)\hat{u}(\lambda, \mu).$$

Here $< \lambda, \partial >= \sum_{i=1}^n \lambda_i \partial_i$. We know that $\hat{u}(\lambda, \mu)$ is a polynomial for all λ in $\mathbf{C}^n$, and μ in $\mathbf{C}$. For a fixed pair λ, μ let

$$\hat{u}(\lambda, \mu)(x, t) = a_N(x)t^N + a_{N-1}(x)t^{N-1} + \cdots + a_0(x),$$

131

where a_k is a polynomial $(k = 0, 1, \ldots, N)$ and $a_N \neq 0$. Substitution and comparing the coefficients gives

$$a_{k+1}(x) = \frac{a^2}{k+1}(\Delta + 2 < \lambda, \partial >)a_k(x).$$

Obviously $a_0(x) = \hat{u}_0(\lambda)(x)$, and hence

$$a_k(x) = \frac{a^{2k}}{k!}(\Delta + 2 < \lambda, \partial >)^k \hat{u}_0(\lambda)(x)$$

holds for $k = 0, 1, \ldots, N$. Here N denotes the smallest nonnegative integer for which $a_N \neq 0$, $a_{N+1} = 0$. The existence of such N follows from the fact that $\hat{u}_0(\lambda)$ is a polynomial. (We supposed here that $\hat{u}_0(\lambda) \neq 0$.) Using the inversion formula (Theorem 7.3) we get

THEOREM 15.8. *Let n be a positive integer and $u_0 : \mathbf{R}^n \to \mathbf{C}$ an exponential polynomial. Then the unique solution of the Cauchy-problem (15.5) can be written in the form*

$$u(x,t) = \sum_{\lambda \in \mathbf{C}^n} \sum_{N=0}^{\infty} \frac{[a^2(\Delta + 2 < \lambda, \partial >)]^N}{N!} \hat{u}_0(\lambda)(x) t^N e^{<\lambda, x> + a^2 <\lambda, \lambda> t}$$

for all x in $\mathbf{R}^n$ and t in $\mathbf{R}$.

A straightforward extension of the above result is

THEOREM 15.9. *Let n be a positive integer and $u_0 : \mathbf{R}^n \to \mathbf{C}$ an exponential polynomial and $t_0 \in \mathbf{R}$. Then the unique solution of the Cauchy-problem*

$$(15.6) \qquad \begin{cases} \partial_t u = a^2 \Delta u, \\ u(x, t_0) = u_0(x) \end{cases}$$

can be written in the form

$$u(x,t) = \sum_{\lambda \in \mathbf{C}^n} \sum_{N=0}^{\infty} \frac{[a^2(\Delta + 2 < \lambda, \partial >)]^N}{N!} \hat{u}_0(\lambda)(x)(t - t_0)^N \times$$

$$\times e^{<\lambda, x> + a^2 <\lambda, \lambda>(t - t_0)}$$

132

for all x in $\mathbf{R}^n$ and t in $\mathbf{R}$.

The next step is to solve the inhomogeneous Cauchy-problem

(15.7)
$$\begin{cases} \partial_t u = a^2 \Delta u + f(x,t), \\ u(x,0) = u_0(x), \end{cases}$$

which can be reduced by a standard procedure to Cauchy-problems of the type (15.5) and (15.6). Then the following theorem is a consequence of Theorems 15.8 and 15.9.

THEOREM 15.10. *Let n be a positive integer and let $f : \mathbf{R}^n \times \mathbf{R} \to \mathbf{C}$, $u_0 : \mathbf{R}^n \to \mathbf{C}$ be exponential polynomials. Then the unique solution of the Cauchy-problem (15.7) can be written in the form $u = u_1 + u_2$, where*

$$u_1(x,t) = \sum_{\lambda \in \mathbf{C}^n} \sum_{N=0}^{\infty} \frac{[a^2(\Delta + 2 < \lambda, \partial >)]^N}{N!} \hat{u}_0(\lambda)(x) t^N e^{<\lambda,x> + a^2 <\lambda,\lambda> t},$$

$$u_2(x,t) = \sum_{\lambda \in \mathbf{C}^n} \sum_{\mu \in \mathbf{C}} \sum_{k=0}^{\infty} \sum_{N=0}^{\infty} (-1)^k \frac{[a^2(\Delta + 2 < \lambda, \partial >)]^N}{N!k!} \times$$

$$\times \partial_{n+1}^k \hat{f}(\lambda,\mu)(x,t) I_{N+k}(a^2 < \lambda, \lambda > -\mu) e^{<\lambda,x> + \mu t}$$

for all x in $\mathbf{R}^n$ and t in $\mathbf{R}$, where $I_k(0) = \frac{t^{k+1}}{k+1}$ and for $\alpha \neq 0$

$$I_k(\alpha) = e^{\alpha t} \sum_{j=0}^{k} (-1)^j \frac{M!}{(M-j)!} \frac{t^{M-j}}{\alpha^{j+1}} - (-1)^M \frac{M!}{\alpha^{M+1}}$$

for $k = 0, 1, \ldots$.

The previous results can easily be extended for evolution equations of more general type, as for instance for the Cauchy-problem

(15.8)
$$\begin{cases} \partial_t u = P(\partial)u + f(x,t), \\ u(x,0) = u_0(x), \end{cases}$$

where $P : \mathbf{R}^n \to \mathbf{C}$ is a polynomial in n variables, $\partial = (\partial_1, \partial_2, \ldots, \partial_n)$, $\partial_t = \partial_{n+1}$, further $f : \mathbf{R}^n \times \mathbf{R} \to \mathbf{C}$, $u_0 : \mathbf{R}^n \to \mathbf{C}$ are exponential polynomials. By

133

the above method we can produce an exponential polynomial solution, which is the only solution if uniqueness is guaranteed. As special cases we get explicit solutions of the Cauchy-problem for the Schrödinger, or biharmonic, or other equations.

THEOREM 15.11. *Let n be a positive integer and let $f : \mathbf{R}^n \times \mathbf{R} \to \mathbf{C}$, $u_0 : \mathbf{R}^n \to \mathbf{C}$ be exponential polynomials. Then the exponential polynomial $u = u_1 + u_2$, where*

$$u_1(x,t) = \sum_{\lambda \in \mathbf{C}^n} \sum_{N=0}^{\infty} \frac{[P(\partial + \lambda) - P(\lambda)]^N}{N!} \hat{u}_0(\lambda)(x) t^N e^{<\lambda,x>+P(\lambda)t},$$

$$u_2(x,t) = \sum_{\lambda \in \mathbf{C}^n} \sum_{\mu \in \mathbf{C}} \sum_{k=0}^{\infty} \sum_{N=0}^{\infty} (-1)^k \frac{[P(\partial + \lambda) - P(\lambda)]^N}{N!k!} \times$$

$$\times \partial_{n+1}^k \hat{f}(\lambda,\mu)(x,t) I_{N+k}(P(\lambda) - \mu) e^{<\lambda,x>+\mu t}$$

for all x in $\mathbf{R}^n$ and t in $\mathbf{R}$, is a solution of Cauchy-problem (15.8).

EXAMPLE 15.12.

We solve the following Cauchy-problem

$$\begin{cases} \partial_t u & = i\Delta u + x \cos t - y^2 \sin t, \\ u(x,y,0) & = x^2 + y^2. \end{cases}$$

Using the above notations we have here

$$P(\lambda_1, \lambda_2) = i(\lambda_1^2 + \lambda_2^2),$$

$$\hat{u}_0(\lambda_1, \lambda_2)(x,y) = \begin{cases} x^2 + y^2 & for \quad \lambda_1 = \lambda_2 = 0, \\ 0 & otherwise, \end{cases}$$

$$\hat{f}(\lambda_1, \lambda_2, \mu)(x,y,t) = \begin{cases} \frac{x}{2} - \frac{y^2}{2i} & for \quad \lambda_1 = \lambda_2 = 0, \mu = i, \\ \frac{x}{2} + \frac{y^2}{2i} & for \quad \lambda_1 = \lambda_2 = 0, \mu = -i, \\ 0 & otherwise. \end{cases}$$

Finally,

$$I_0(\pm i) = \frac{e^{\pm i} - 1}{\pm i}, \qquad I_1(\pm i) = \frac{t e^{\pm it}}{\pm i} + e^{\pm it} - 1,$$

and hence we obtain by Theorem 15.11

$$u_1(x, y, t) = x^2 + y^2 + 4it,$$

$$u_2(x, y, t) = x \sin t + y^2 (\cos t - 1) - 2it + 2i \sin t,$$

and the solution is

$$u(x, y, t) = x \sin t + x^2 + y^2 \cos t + 2i(t + \sin t).$$

REFERENCES 15.13.

For references concerning the above results see [SZÉ7], [SZÉ16], [SZÉ17].

16. Noncommutative applications

Although the methods and results of the previous sections depend heavily on the commutative structure of the underlying groups or semigroups, in some cases they can be applied also in the noncommutative situation. The general idea is to use factorizations through the natural homomorphism $G \mapsto G/C$, where C denotes the commutator subgroup of G. This possibility has been mentioned in special cases in [CHU], [SZÉ20]. Here we give some hints for more general cases, the details being left to the reader.

Let G be any group and let C denote the commutator subgroup of G, that is, the subgroup generated by all elements of the form $xyx^{-1}y^{-1}$ with x, y in G. Then C is a normal subgroup and G/C is commutative. If H is a set, then obviously any function $F : G/C \to H$ induces naturally a function $f = F \circ \Phi$ from G into H, where $\Phi : G \to G/C$ is the natural homomorphism. Moreover, f is constant on the cosets of C. Conversely, any function $f : G \to H$, which is constant on the cosets of C can be factored through Φ, that is, can be written in the form $f = F \circ \Phi$ with some function $F : G/C \to H$. In this case we may consider f heuristically as a function rather on G/C than on G. The advantage of this factorization process is that it is linear (supposing that H has some linear structure), and it is "translation-covariant", that is, $f = F \circ \Phi$ implies

$$\tau_{\Phi(y)} F \circ \Phi = \tau_y f$$

for any y in G. Hence many functional equations of the type we have dealt with are preserved, that is, if f satisfies an equation of that type, then F satisfies a similar one - on the commutative group G/C. Solving it for F we obtain the desired representation of f in the form $f = F \circ \Phi$. From this point of view the following lemma is useful.

LEMMA 16.1. *Let G be a group and H a set. The function $f : G \to H$ is constant on the cosets of the commutator subgroup of G if and only if it satisfies*

(16.1) $$f(xyz) = f(yxz)$$

136

for all x, y, z in G.

PROOF: Suppose that f satisfies the given condition and let C denote the commutator subgroup of G. If a, b belong to the same coset of C, then $ab^{-1} \in C$, that is,

$$a = x_1 y_1 x_1^{-1} y_1^{-1} x_2 y_2 x_2^{-1} y_2^{-1} \ldots x_k y_k x_k^{-1} y_k^{-1} b$$

with some x_i, y_i in G $(i = 1, 2, \ldots, k)$. Then we have

$$f(a) = f(x_1 y_1 x_1^{-1} y_1^{-1} x_2 y_2 x_2^{-1} y_2^{-1} \ldots x_k y_k x_k^{-1} y_k^{-1} b) =$$

$$= f(y_1 x_1 x_1^{-1} y_1^{-1} x_2 y_2 x_2^{-1} y_2^{-1} \ldots x_k y_k x_k^{-1} y_k^{-1} b) =$$

$$= f(x_2 y_2 x_2^{-1} y_2^{-1} \ldots x_k y_k x_k^{-1} y_k^{-1} b) = \cdots = f(b),$$

that is, f is constant on the cosets of C. Conversely, if f is constant on the cosets of C, then

$$f(xyz) = f(xyx^{-1}y^{-1}yxz) = f(yxz)$$

for all x, y, z in G, hence our lemma is proved.

Equation (16.1) as an additional assumption has been used by several authors (see [ACZ17], [KNP], [SZÉ20]) without mentioning that in this case the problem can be reduced to the commutative case. We note that different equations show a different behavior concerning condition (16.1): in some cases (16.1) is a consequence of the equation itself, and in other cases it must be additionally assumed. Another feature of (16.1) is that although it obviously implies

$$f(xy) = f(yx)$$

for all x, y in G, but the converse implication is false in general.

Now we consider some examples. We note, that in the noncommutative case the translation operator τ_y is meant to denote the right translation, that is

$$\tau_y f(x) = f(xy)$$

holds for all x, y, and respectively, the difference operators $\Delta_{y_1, y_2, \ldots, y_n}$ are built up by right translations. Additive and multi-additive functions on noncommutative groups have the obvious meaning.

137

THEOREM 16.2. *Let G be a group, S a linear space over the rationals and n a positive integer. Let $f : G \to S$ be a function satisfying (16.1) and (9.2). Then*

$$f = \sum_{k=0}^{n} A_k^*,$$

where $A_k : G^k \to S$ is a k-additive symmetric function satisfying (16.1) in each variable $(k = 0, 1, \ldots, n)$.

PROOF: Using Lemma 16.1 we can factor f through the natural homomorphism Φ of G onto G/C in the form

$$f = F \circ \Phi$$

with some $F : G/C \to S$. Then we have

$$\Delta_v^{n+1} F(u) = \Delta_y^{n+1} f(x)$$

for all $u = \Phi(x)$, $v = \Phi(y)$ with x, y in G. Using the surjectivity of Φ we can apply Theorem 9.3 to get the statement.

We note that in general, (16.1) is not a consequence of (9.2) (see [NG]).

Our next example is the Levi-Civitá equation (10.1). Concerning its non-degenerate solutions one sees, that if f satisfies (16.1), then so do g_i, h_i for $i = 1, 2, \ldots, n$. It is the case for instance, if

$$g_i(xy) = g_i(yx)$$

holds for $i = 1, 2, \ldots, n$ and for all x, y in G. Then by factorization we have that the functions

$$f = F \circ \Phi,$$

$$g_i = G_i \circ \Phi,$$

$$h_i = H_i \circ \Phi$$

$(i = 1, 2, \ldots, n)$ satisfy (10.1) on the commutative group G/C, hence our results in Section 10 can be applied. As a special case we mention the equation

$$f(xy) = f(x)g(y) + g(x)f(y) + h(x)h(y)$$

(see [CHU]). Here the equation implies that f, g, h are constant on the cosets of the commutator subgroup, as it is shown in [CHU], hence Theorem 10.4 can be applied directly to get the solutions as listed in [CHU].

Concerning the equations treated in Section 11 we mention here the d'Alembert equation on noncommutative groups, solved in [KNP], where the condition (16.1) is explicitly assumed, hence Theorem 12.7 can be applied. Similar arguments can be used concerning equations treated in [ACZ17].

The above mentioned methods can also be applied for equations considered in Section 13. The details are left to the reader.

REFERENCES 16.3.

For references concerning this section see [ACZ17], [CHU], [KNP], [NG], [SZÉ20], [SZÉ24].

APPENDIX

The aim of this book was to present a unified theory of convolution type functional equations and to illustrate the application possibilities of the results of spectral analysis and spectral synthesis to functional equations. Of course, this presentation is far from being complete, but it indicates that the method may serve as a basis for the study of a wide class of functional equations, including several classical ones, and it is also able to produce new results. Nevertheless, some possible extensions of the method should be investigated. Here we collect some directions, in which further results would be interesting.

It would be interesting to know if, and to what extent functional equations with some restrictions can be solved by using spectral synthesis. It seems that the use of so-called uniqueness sets makes it possible to prove results analogous to the existing ones in case of the classical equations also in the case, if one of the variables varies not in the whole group, but it is subjected to some restrictions, for instance, it varies in a "smaller" set. This smaller set is called a uniqueness set, if it has the property, that any two exponential polynomials, which are equal on this set, are identical. For instance, in a topological abelian group, which is generated by every neighborhood of zero any nonvoid open set is a uniqueness set. Of course, an explicit characterization of uniqueness sets would also be interesting.

Another extension towards local functional equations would be desirable. Obviously here some extension of the basic spectral synthesis results for local translation invariant subspaces is needed. Maybe an appropriate notion of local exponential polynomials using local exponentials and local additive functions could serve as a useful tool.

As several examples of the classical equations show, some of the existing results remain valid also in the case, if the underlying structure is a commutative semigroup only (polynomial equation, Levi-Cività equation, etc.). It would be interesting to know if some results of spectral synthesis can be extended for commutative semigroups. Unfortunately, on semigroups a basic tool, the convolution fails to exist, or should be substituted by some "one-sided" convolution. But the other fundamental notions, like polynomials, exponentials, exponential polynomials all have their obvious meaning.

Another interesting problem is the extension of the results to noncommutative groups. Here the methods of the noncommutative harmonic analysis should be used, as e.g. finite dimensional representations, etc. For the noncommutative case an appropriate substitute of the exponential monomials is needed, which may serve as basic functions for possible approximation processes.

Finally, let's collect here some functional equations whose general solutions have not been completely characterized yet.

The n-dimensional generalization of the wave equation:

$$[\prod_{i=1}^{n}(X_i^t - 1)]f = 0,$$

or

$$[\prod_{i=1}^{n}(X_i^t - X_i^{-t})]f = 0;$$

the respective sum-form equation:

$$[\sum_{i=1}^{n}(X_i^t - X_i^{-t})]f = 0;$$

the n-dimensional octahedron equation:

$$[\sum_{i=1}^{n}(X_i^t + X_i^{-t})]f = 2nf;$$

and the n-dimensional octahedron-cube equation:

$$[\prod_{i=1}^{n}(X_i^t + X_i^{-t})]f = [\sum_{i=1}^{n}(X_i^t + X_i^{-t})]f + (2^n - 2n)f$$

(which is equivalent to the octahedron - and to the cube - equation for $n \leq 3$).

A possible generalization (and solution) of the following equations for n variables would be interesting too:

$$(X^t + 1)(Y^t - 1)f = 0,$$

$$(X^t + X^{-t})f = (Y^t + Y^{-t})f,$$

$$(X^t - X^{-t})f = (Y^t - Y^{-t})f.$$

141

Index

143

REFERENCES

[ACZ1] J.Aczél, L.Jánossy, A.Rényi, On composed Poisson distributions, *Acta Math. Acad. Sci. Hung.*, **1**(1950) 209-224.

[ACZ2] J.Aczél, On composed Poisson distributions III., *Acta Math. Acad. Sci. Hung.*, **3**(1952) 219-224.

[ACZ3] J.Aczél, Über Additions- und Subtraktionstheoreme, *Publ. Math. Debrecen*, **4**(1955/56) 325-333.

[ACZ4] J.Aczél, Miszellen über Funktionalgleichungen I., *Math. Nachr.*, **19**(1958) 88-99.

[ACZ5] J.Aczél, Sur une classe d'équations fonctionelles bilinéaires à plusieurs fonctions inconnues, *Publ. Fac. Électrotechn. Univ. Belgrade, Ser. Math.- Phys.*, **No.61-64**(1961) 12-20.

[ACZ6] J.Aczél, Some unsolved problems in the theory of functional equations I., *Arch. Math.*, **15**(1964) 435-444.

[ACZ7] J.Aczél,"Lectures on Functional Equations and Their Applications", Academic Press, New York and London, 1966.

[ACZ8] J.Aczél, H.Haruki, M.A.McKiernan, G.N.Sakovič, General and regular solutions of functional equations characterizing harmonic polynomials, *Aequationes Math.*, **1**(1968) 37-53.

[ACZ9] J.Aczél, G.Vranceanu, Equations fonctionelles liées aux groupes lineaires commutatifs, *Colloq. Math.*, **26**(1972) 371-383.

[ACZ10] J.Aczél, Functions of Binomial Type Mapping Groupoids into Rings, *Math. Zeitschrift*, **154**(1977) 115-124.

[ACZ11] J.Aczél, Some good and bad characters I have known and where they led, in Harmonic analysis and functional equations, 1980 Seminar on Harmonic Analysis, Canad. Math. Soc. Conf. Proc., **1**, (Amer. Math. Soc. Providence, R.I.)(1981) 177-187.

[ACZ12] J.Aczél, J.K.Chung, Integrable solutions of functional equations of a general type, *Studia Sci. Math. Hung.*, **17**(1982/84) 51-67.

[ACZ13] J.Aczél, Some unsolved problems in the theory of functional equations II., *Aequationes Math.*, **26**(1983-84) 255-260.

[ACZ15] J.Aczél, D.Reidel (editors),"Functional Equations: History, Applications and Theory", Publishing Company, Dordrecht, Boston, Lancaster, 1984.

[ACZ16] J.Aczél, On history, applications and theory of functional equations, in "Functional Equations: History, Applications and Theory", Publishing

Company, Dordrecht, Boston, Lancaster, 1984, 3-12.

[ACZ17] J.Aczél, Symmetric second differences in product form on groups, Talk given at the 25th ISFE, Hamburg-Rissen, Germany, 1987. (Report of Meeting, *Aequationes Math.*, **35**(1988) 82-124, p.83.)

[ACZ18] J.Aczél, J.Dhombres,"Functional Equations Containing Several Variables", Encyclopedia of Mathematics and its Applications, **30**, Cambridge Univ.Press, 1988.

[ALE] A.Alexiewicz,W.Orlicz, Analytic Operations in Real Banach Spaces, *Studia Math.*, **14**(1953) 57-78.

[ALR1] A.M.Al-Rashed, H.T.Hemdan, On four-dimensional functional equations, *Journal of the College of Science*, **17**(1986) 243-251.

[ALR2] A.M.Al-Rashed, H.T.Hemdan, Solution of the generalized cube and octahedron functional equations, *Util. Math.*, **33**(1988) 123-135.

[ANG1] T.Angheluta, Sur une équation fonctionelle caractérisant les polynómes, *Bull. Soc. Sci. Cluj*, **6**(1931) 139-145.

[ANG2] T.Angheluta, Sur une équation fonctionelle caractérisant les polynómes, *Mathematica (Cluj)*, **6**(1932) 1-7.

[ANS] P.M.Anselone, J.Korevaar, Translation invariant subspaces of finite dimension, *Proc. Amer. Math. Soc.*, **15**(1964) 747-752.

[BAK1] J.A.Baker, An analogue of the wave equation and certain related equations, *Canadian Math. Bull.*, **12**(1969) 837-846.

[BAK2] J.A.Baker, Regularity Properties of Functional Equations, *Aequationes Math.*, **6**(1971) 243-247.

[BAK3] J.A.Baker, On the functional equation $f(x)g(y) = \sum_{i=1}^{n} a_i(x)b_i(y)$, *Aequationes Math.*, **11**(1974) 154-162.

[BAK4] J.A.Baker, On the functional equation $f(x)g(y) = p(x+y)q(\frac{x}{y})$, *Aequationes Math.*, **14**(1976) 493-506.

[BAN] S.Banach, Sur l'équation fonctionelle $f(x+y) = f(x)+f(y)$, *Fund. Math.*, **1**(1930) 123-124.

[BEL] E.T.Bell, Exponential polynomials, *Annals of Math.*, **35**(1934) 258-277.

[BEN] J.J.Benedetto,"Spectral Synthesis", Academic Press, New York, London, San Francisco, 1975.

[BEU] A.Beurling, On the spectral synthesis of bounded functions, *Acta Math.*, **81**(1949) 225-238.

[BOU1] N.Bourbaki,"Intégration", Hermann, Paris, 1952,1956.

[BOU2] N.Bourbaki,"General Topology", Addison-Wesley Publishing Company Reading, Massachusettes - Palo Alto - London - Don Mills, Ontario, 1966.

[CAR1] F.W.Carroll,"Doctoral Thesis", Purdue University, Lafayette, Ind., 1959.

[CAR2] F.W.Carroll, A difference property for polynomials and exponential polynomials on abelian locally compact groups, *Trans. Amer. Math. Soc.*, **114**(1965) 147-155.

[CAU] A.L.Cauchy,"Cours d'Analyse", Oeuvres, Ser.2., **3**(1987) 106-113.

[CHU] J.K.Chung, Pl.Kannappan, C.T.Ng, A generalization of a cosine-sine functional equation on groups, *Linear Algebra Appl.*, **66**(1985) 259-277.

[COI] I.Corovei, The cosine functional equation for nilpotent groups, *Aequationes Math.*, **15**(1977) 99-106.

[COR1] L.Corwin, A"Functional Equation" for Measures and a Generalization of Gaussian Measures, *Bull. Amer. Math. Soc.*, **75**(1969) 829-832.

[COR2] L.Corwin, Generalized Gaussian Measures and a"Functional Equation" I., *J. Funct. Anal.*, **5**(1970) 412-427.

[COR3] L.Corwin, Generalized Gaussian Measures and a"Functional Equation" II., *J. Funct. Anal.*, **6**(1970) 481-505.

[DAR] Z.Daróczy, Elementare Lösung einer mehrere unbekannte Funktionen enthaltenden Funktionalgleichung, *Publ. Math. Debrecen*, **8**(1961) 160-168.

[DIE] J.Dieudonné,"Foundations of Modern Analysis", Academic Press, London, 1960.

[DIT] V.Ditkin, On the structure of ideals in certain normed rings, *Uchenye Zapiski Moskov. Gos. Univ. Matematika*, **30**(1939) 83-130.

[DJO1] D.Z.Djokovič, A theorem on semigroups of linear operators, *Publ. de l'Institut Mathématique de Beograd, Nouvelle Série*, **3(17)**(1963) 129-130.

[DJO2] D.Z.Djokovic, Triangle functional equation and its application, *Univ. Beogr. Publ. Elektrot. Fak. Ser. Mat. Fiz.* (1967) 181-196.

[DJO3] D.Z.Djokovic, A representation theorem for $(X_1 - 1)(X_2 - 2)\ldots(X_n - 1)$ and its applications, *Ann. Polon. Math.* **22**(1969) 189-198.

[ECS1] I.Ecsedi, Az $f(ax + by)g(cx + dy) = h(x)k(y)$ függvényegyenlet nem folytonos megoldásainak egy osztályáról, *MTA III. Oszt. Közl.*, **22**(1974) 3-10.

[ECS2] I.Ecsedi, On a functional equation which contains a real-valued function of vector variable, *Period. Math. Hung.*, **5**(1974) 333-342.

[EDG] G.A.Edgar, J.M.Rosenblatt, Difference equations over locally compact abelian groups, *Trans. Amer. Math. Soc.* **253**(1979) 257-289.

[EHR1] L.Ehrenpreis, Mean periodic functions, *Amer. J. Math.*, **77**(1955) 293-328.

[EHR2] L.Ehrenpreis, Appendix to"Mean periodic functions", *Amer. J. Math.*, **77**(1955) 731 - 733.

[ELL1] R.J.Elliot, Some results in spectral synthesis, *Proc. Cambridge Phil. Soc.*, **61**(1965) 395-424.

[ELL2] R.J.Elliot, Two notes on spectral synthesis for discrete Abelian groups, *Proc. Cambridge Phil. Soc.*, **61**(1965) 617-620.

[ENG] M.Engert, Finite dimensional translation invariant subspaces, *Pacific J. Math.*, **32**(1970) 333-343.

[ETG] L.Etigson, Equivalence of 'Cube' and 'Octahedron' functional equations, *Aequationes Math.*, **10**(1974) 50-65.

[FED] H.Federer,"Geometric Measure Theory", Springer Verlag, Berlin, Heidelberg, New York, 1969.

[FLE] T.M.Flett, Continuous solutions of the functional equation $f(x + y) + f(x - y) = 2f(x)f(y)$, *Amer. Math. Monthly*, **70**(1963) 392-397.

[FRE1] M.Fréchet, Toute fonctionelle continue est développable en série de fonctionelles d'ordre entier, *Compt. Rend. Acad. Sci. (Paris)*, **148**(1909) 155-156.

[FRE2] M.Fréchet, Une définition fonctionelle des polynômes, *Nouv. Ann.*, **49**(1909) 145-162.

[FRE3] M.Fréchet, Pri la funkcia equacio $f(x + y) = f(x) + f(y)$, *L'Enseignement Math.*, **15**(1913) 390-393.

[FRE4] M.Fréchet, Les polynômes abstraits, *Journal de Math.*, **8**(1929).

[GAJ1] Z.Gajda, On some properties of Hamel bases connected with the continuity of polynomial functions, *Aequationes Math.*, **27**(1984) 57-75.

[GAJ2] Z.Gajda, Additive and convex functions in linear topological spaces, *Aequationes Math.*, **27**(1984) 214-219.

[GAJ3] Z.Gajda, Christensen measurable solutions of generalized Cauchy functional equations, *Aequationes Math.*, **31** (1986) 147-158.

[GAJ4] Z.Gajda, A solution of a problem of J.Schwaiger, *Aequationes Math.*, **32**(1987) 38-44.

[GAJ5] Z.Gajda, A characterization of exponential polynomials by a class of functional equations, *Publ. Math. Debrecen*, **35(1-2)**(1988) 51-63.

[GAJ6] Z.Gajda, Christensen measurability of polynomial functions and convex functions of higher orders, *Ann. Polon. Math.*, **47(1)**(1986) 25-40.

[GER1] R.Ger, On some properties of polynomial functions, *Ann. Polon. Math.*, **25**(1971) 195-203.

[GER2] R.Ger, Mazur's criterion for continuity of convex functionals, Talk given at the 25th ISFE in Hamburg-Rissen, Germany, 1987. (Report of Meeting, *Aequationes Math.*, **35**(1988) 82-124).

[GHE1] M.Ghermanescu, Solutions mesurables de certaines équations fonctionelles linéaires à plusieurs variables, *Bull. Sci. École Polytech. Timişoara*, **13**(1948) 18-37.

[GHE2] M.Ghermanescu, Caractérisation fonctionelle des fonctions trigonométriques, *Bull. Inst. Polyt. Iaşi*, **4**(1949) 362-368.

[GHE3] M.Ghermanescu,"Ecuatii Functionale", Bucureşti, 1960.

[GHU] S.G.Ghurye, I.Olkin, A characterization of the multivariate normal distribution, *Amer. Math. Statist.*, **33**(1962) 533-541.

[GIL1] J.E.Gilbert, Two notes on spectral synthesis, *Math. Proc. Cambridge Phil. Soc.*, **60**(1966) 618-623.

[GIL2] J.E.Gilbert, Spectral synthesis problems for invariant subspaces on groups, *Amer. J. Math.*, **88**(1966) 626-635.

[GIR1] D.Girod, J.H.B.Kemperman, On the functional equation $\sum_{j=0}^{n} a_j f(x + T_j y) = 0$, *Aequationes Math.*, **3**(1970) 230.

[GIR2] D.Girod, On the Functional Equation $\Delta_{T_1 y} \Delta_{T_2 y} f = 0$, *Aequationes Math.*, **9** (1973) 157-164.

[HAJ] O.Hájek, Sur les équations fonctionelles des fonctions trigonométriques (Russian), *Czech. Math. J.*, **5(80)**(1955) 432-434.

[HLL] R.L.Hall, On single-product functions with rotational symmetry, *Aequationes Math.*, **8**(1972) 291-296.

[HAL] P.R.Halmos,"Measure Theory", Van Nostrand, New York, 1950.

[HAM] G.Hamel, Eine Basis aller Zahlen und die unstetigen Lösungen der Funktionalgleichung $f(x + y) = f(x) + f(y)$, *Math. Ann.*, **60**(1905) 459-462.

[HAS1] S.Haruki,,"Studies on Functional Equations", Master Thesis, University of Waterloo, Waterloo, Ontario, Canada, 1972.

[HAS2] S.Haruki, A note on a pentomino functional equation, *Ann. Polon. Math.*, **27**(1973) 129-131.

[HAS3] S.Haruki, Note on the equation $(\Delta_{x,t}^2 - \Delta_{y,t}^2)f = 0$, *Ann. Polon. Math.*, **21**(1980) 49-52.

[HAS4] S.Haruki, On the theorem of S.Kakutani-M.Nagumo and J.L.Walsh for the mean value property of harmonic and complex polynomials, *Pacific J. Math.*, **94**(1981) 113-123.

[HAS5] S.Haruki, On the general solution of the triangle mean value equation, *Aequationes Math.*, **25**(1982) 209-215.

[HHA1] H.Haruki, On a 'Cube' functional equation, *Aequationes Math.*, **3**(1969) 156-159.

[HHA2] H.Haruki, On the functional equation $f(x + t, y) + f(x - t, y) = f(x, y + t) + f(x, y - t)$, *Aequationes Math.*, **5**(1970) 118-119.

[HEL1] H.Helson, Spectral synthesis of bounded functions, *Ark.Mat.*, **1(34)** (1951) 497-502.

[HEL2] H.Helson,"Harmonic Analysis", Addison-Wesley Publishing Company, London, Amsterdam, Sydney, Toyko, 1983.

[HEM] H.T.Hemdan, Solution of the 'Cube' functional equation in 'Trilinear Coefficients', *Canadian Math. Bull.*, **19** (1976) 181-191.

[HEW] E.Hewitt, K.Ross,"Abstract Harmonic Analysis I,II", Springer Verlag, Berlin, 1963.

[HOS1] M.Hosszú,"Algebrai rendszereken értelmezett függvényegyenletek", Dissertation, Miskolc, 1963.

[HOS2] M.Hosszú, On the Fréchet's functional equation, *Bul. Inst. Polit. Iaşi*, **10**(1964) 27-28.

[HOS3] M.Hosszú, Some remarks on the cosine functional equation, *Publ. Math. Debrecen*, **16**(1969) 93-98.

[ING] M.H.Ingraham, Solutions of certain functional equations relative to a general linear set, *Trans. Amer. Math. Soc.*, **28**(1926) 287-300.

[ILS1] D.Ilse, Bemerkungen zu den Additions- und Multiplikationstheoremen einiger elementarer Funktionen, *Math. Schule*, **10**(1972) 193-205.

[ILS2] D.Ilse, Über rationale Additionstheoreme und verwandte Funktionalgleichungen, *Math. Schule*, **11**(1973) 392-403.

[ILS3] D.Ilse, Zur funktionalen Charakterisierung der Winkelfunktionen, *Math. Schule*, **14**(1976) 87-96.

[JAR1] A.Járai, On measurable solutions of functional equations, *Publ. Math. Debrecen*, **26**(1979) 17-35.

[JAR2] A.Járai, On regular solutions of functional equations, *Aequationes Math.*, **30**(1986) 21-54.

[JAR3] A.Járai, A remark to a paper of J.Aczél and J. K. Chung, *Studia Sci. Math. Hung.*, **19(2-4)**(1984) 273-274.

[KAG] A.M.Kagan, Yu.V.Linnik, C.R.C.Rao," Characterization Problems in Mathematical Statistics", John Wiley and Sons, New York, 1973.

[KAH1] J.P.Kahane, Sur quelques problémes d'unicité et prolongement relatifs aux fonctions approchables par des sommes d'exponentielles, *Ann. Inst. Fourier (Grenoble)*, **5** (1953-54) 39-130.

[KAH2] J.P.Kahane, Sur les fonctions moyenne-périodiques bornées, *Ann. Inst. Fourier (Grenoble)*,**7**(1957) 293-314.

[KAN] D.Kannan, Pl.Kannappan, On a characterization of Gaussian measures in a Hilbert space, *Notices Amer. Math. Soc.*, **23**(1976) A-549.

[KAP] I.Kaplansky, Primary ideals in group algebras, *Proc. Nat. Acad. Sci.*, **35**(1949) 133-136.

[KNP] Pl.Kannappan, The functional equation $f(xy) + f(xy^{-1}) = 2f(x)f(y)$ for groups, *Proc. Amer. Math. Soc.*, **19**(1968) 69-74.

[KEM1] J.H.B.Kemperman, A general functional equation, *Trans. Amer. Math. Soc.*, **86**(1957) 28-56.

[KEM2] J.H.B.Kemperman, A property of exponential polynomials, *Bull. Am. Math. Soc.*, **63**(1957) 33.

[KEM3] J.H.B.Kemperman, On a generalized difference property, *Aequationes Math.*, **3**(1970) 231.

[KEM4] J.H.B.Kemperman, On exponential polynomials, (unpublished manuscript).

[KEN] K.McKennon, B.Dearden, Functional equations for polynomials, *Proc. Amer. Math. Soc.*, **63(1)**(1977) 23-28.

[KES] H.Kestelman, On the functional equation $f(x + y) = f(x) + f(y)$, *Fund. Math.*, **34**(1947) 144-147.

[KHA1] C.G.Khatri, C.R.Rao, Solutions to some functional equations and their applications to characterization of probability distributions, *Sankhya, Ser.A*, **30**(1968) 167-180.

[KHA2] C.G.Khatri, C.R.Rao, Functional equations and characterization of probability laws through linear functions of random variables, *J. Multiv. Anal.*, **2**(1972) 162-173.

[KCZ1] M.Kuczma, A.Zajtz, Über die multiplikative Cauchysche Funktionalgleichung für Matrizen dritter Ordnung, *Arch. Math.*, **15**(1964) 136-143.

[KCZ2] M.Kuczma, A.Zajtz, On the form of real solutions of the matrix functional equation $\Phi(xy) = \Phi(x)\Phi(y)$ for nonsingular matrix Φ, *Publ. Math. Debrecen*, **13**(1966) 257-262.

[KCZ3] M.Kuczma, A.Zajtz, Quelques remarques sur l'équation fonctionelle matricielle de Cauchy, *Colloq. Math.*, **18**(1967) 159-168.

[KCZ4] M.Kuczma,"An Introduction to the Theory of Functional Equations and Inequalities", Państwowe Wydawnictwo Naukowe, Uniwersytet Śląski, Warszawa, Kraków, Katowice, 1985.

[KUR1] S.Kurepa, Semigroups of linear transformations in n-dimensional vector space, *Glasnik Mat. Fiz.*, **13**(1958) 3-32.

[KUR2] S.Kurepa, On some functional equations in Banach spaces, *Stud. Math. Appl.*, **19**(1960) 147-158.

[KUR3] S.Kurepa, On the functional equation $f(x + y)f(x - y) = f^2(x) - f^2(y)$, *Ann. Polon. Math.*, **10**(1961) 1-5.

[KUR4] S.Kurepa, A property of a set of positive measure and its application, *J. Math. Soc. Japan*, **13**(1961) 13-19.

[LAB] I.Labuda, R.D.Mauldin, Problem 24. of the"Scottish Book" concerning additive functionals, *Colloq. Math.*, **48**(1984) 89-91.

[LAI1] P.G.Laird, Entire mean periodic functions, *Canad. J. Math.*, **27**(1975) 805-818.

[LAI2] P.G.Laird, On characterizations of exponential polynomials, *Pacific J. Math.*, **80**(1979) 503-507.

[LAI3] P.G.Laird, A reconsideration of the"three squares problem", *Aequationes Math.*, **21**(1980) 98-104.

[LAI4] P.G.Laird, J.Mills, On systems of linear functional equations, *Aequationes Math.*, **26**(1983) 64-73.

[LAJ1] K.Lajkó,"Több ismeretlen függvényt tartalmazó függvényegyenletekről", Thesis, University of Debrecen, Debrecen, 1969.

[LAJ2] K.Lajkó,"Differencia-tulajdonságok, függvényegyenletek és karakterizációs problémák", Dissertation, University of Gödöllő, Debrecen-Gödöllő, 1978.

[LAJ3] K.Lajkó, Remark to a paper of J.A.Baker, *Aequationes Math.*, **19**(1979) 227-231.

[LAJ4] K.Lajkó, On the functional equation $f(x)g(y) = h(ax + by)k(cx + dy)$, *Periodica Mat. Hung.*, **11**(1980) 187-195.

[LEF] M.Lefranc, L'analyse harmonique dans $\mathbf{Z}^n$, *C. R. Acad. Sci. Paris*, **246**(1958) 1951-1953.

[LES1] J.A.Lester, A canonical form for a system of quadratic functional equations, *Ann. Polon. Math.*, **35**(1976) 105.

[LES2] J.A.Lester, The solution of a sytem of quadratic functional equations, *Ann. Polon. Math.*, **37**(1980) 113-117.

[LEV] T.Levi-Civita, Sulle funzioni che ammettono una formula d'addizione del tipo $f(x + y) = \sum_{i=1}^{n} X_i(x)Y_i(y)$, *Atti Accad. Nazl. Lincei, Rend.*, **(5)22**(1913) 181-183.

[LIJ1] G.Van der Lijn, Les polynômes abstraits, I., *Bull. Sci. Math.*, **(2)64**(1940) 55-80.

[LIJ2] G.Van der Lijn, Les polynômes abstraits, II., *Bull. Sci. Math.*, **(2)64**(1940) 102-112.

[LIJ3] G.Van der Lijn, La definition des polynômes dans les groups abéliens, *Fund. Math.*, **33** (1945) 42-50.

[LIP] Z.Lipecki, On continuity of group homomorphisms, *Colloq. Math.*, **48**(1984) 93-94.

[LOO] L.Loomis,"Introduction to Abstract Harmonic Analysis", van Nostrand, Princeton, Toronto, London, Melbourne, 1953.

[LUK] E.Lukács, A characterization of the normal distribution, *Ann. Math. Stat.*, **13**(1942) 91-93.

[MAA] W.Maak,"Fastperiodische Funktionen", Springer Verlag, 1950.

[MAC] J.A.MacDougall, K.Ozeki, The octahedron equation implies the cube equation: an elementary proof, *Aequationes Math.*, **31**(1986) 243-246.

[MAG] L.J.Magnus, Über die Relationen der Funktionen, welche der Gleichung
$F_1 y\varphi_1 x + F_2 y\varphi_2 x + \ldots \cdots + F_n y\varphi_n x = F_1 x\varphi_1 y + F_2 x\varphi_2 y + \cdots + F_n x\varphi_n y$
genug tun, *J.Reine Angew. Math.*, **5**(1830) 365-373.

[MAK] Gy.Maksa,"Deviációk és differenciák", Thesis, University of Debrecen, Debrecen, 1976.

[MAL] P.Malliavin, Impossibilité de la synthèse spectrale sur les groupes abéliens non compacts, *Inst. des Hautes Etudes Scientifiques, Publications Maths.*, **2**(1959) 61-68.

[MAR] A.Marchaud, Sur les dérivées et sur les différences des fonctions de variables réelles, *J. Math. Pures Appl.*, **(9)6**(1927) 337-435.

[MAU] J.G.Mauldon, Continuous functions satisfying linear recurrence relations, *Quaterly J. Math.*, **15**(1964) 23-31.

[MAZ] S.Mazur, W.Orlicz, Grundlegende Eigenschaften der polynomischen Operationen, I.,II., *Studia Math.*, **5**(1934) 50-68, 179-189.

MCK1] M.A.McKiernan, On vanishing n-th ordered differences and Hamel bases, *Ann. Polon. Math.*, **19**(1967) 331-336.

MCK2] M.A.McKiernan, General solution of quadratic functional equations, *Aequationes Mat.*, **3**(1970).

MCK3] M.A.McKiernan, Boundedness on a set of positive measure and the mean value property characterizes polynomials on a space V^n, *Aequationes Math.*, **4**(1970) 31-36.

MCK4] M.A.McKiernan, The general solution of some functional equations analoguous to the Wave equation, *Aequationes Math.*, **8**(1972) 263-266.

MCK5] M.A.McKiernan, Measurable solutions of quadratic functional equations, *Colloq. Math.*, **35**(1976) 97.

MCK6] M.A.McKiernan, The matrix equation $a(x \circ y) = a(x) + a(x)a(y) + a(y)$, *Aequationes Math.*, **15**(1977) 213-223.

MCK7] M.A.McKiernan, Equations of the form $H(x \circ y) = \sum_i f_i(x)g_i(y)$, *Aequationes Math.*, **16**(1977) 51-58.

[MGM] S.Mandelbrojt, S.Agmon, Une généralisation du théorème taubérien de Wiener, *Acta Sci. Math. Szeged*, **12** (1950) 167-176.

[MGR] B.Malgrange, Sur quelques propriétés des equations des convolution, *C.R.Acad.Sci.Paris*, **238**(1954) 2219-2221.

[MIL] J.Mills,"Systems of Linear Functional Equations", Thesis, University of Wollongong, Wollongong, 1982.

[NG] C.T.Ng, The Jensen equation on groups, Talk given at the 26th ISFE, San Feliu de Guixols, Spain 1988. (Report of Meeting, *Aequationes Math.*, **37**(1989) 57-127., 80).

[NOV] D.Novak, K.McKennon, Exponentials on Locally Compact Abelian Groups, *Proc. Amer. Math. Soc.*, **83**(1981) 307-314.

[OCO1] T.A.O'Connor, A solution of the functional equation $\Phi(x - y) = \sum_1^n a_j(x)\bar{a}_j(y)$ on a locally compact Abelian group, *Aequationes Math.*, **15**(1977) 113.

[OCO2] T.A.O'Connor, A solution of d'Alembert's functional equation on a locally compact Abelian group, *Aequationes Math.*, **15**(1977) 235-238.

[OCO3] T.A.O'Connor, A Solution of the Functional Equation $\varphi(x - y) = \sum_1^n a_j(x)\bar{a}_j(y)$ on a Locally Compact Abelian Group, *Journal of Math. Anal. and Appl.*, **60(1)** (1977) 120-122.

[OST] A.Ostrowski, Mathematische Miszellen, XIV: Über die Funktionalgleichung der Exponentialfunktion und verwandte Funktionalgleichungen, *Jber. Deutsch. Math.- Verein,*, **38** (1929) 54-62.

[PGA1] L.Paganoni, Measurability and continuity for the solutions of a general functional equation, *Aequationes Math.*, **12**(1975) 284-285.

[PGA2] L.Paganoni, On a functional equation concerning affine transformations, *J. Math. Anal. Appl.*, **127**(1987) 475-491.

[PGM] S.Paganoni-Marzegalli, Boundedness and continuity for the solutions of a class of functional equations, *Aequationes Math.*, **14**(1976) 213-214.

[PEN] R.C.Penney,A.L.Rukhin, D'Alembert's functional equation on groups, *Proc. Amer. Math. Soc.*, **77**(1979) 73-80.

[PER] O.Perron, Über Additions- und Subtraktionstheoreme, *Archiew Math. Phys.*, **(3)28**(1919/20) 97-100.

[PEX] H.W.Pexider, Notiz über Funktionaltheoreme, *Monatsh. Math. Phys.*, **14**(1903) 293-301.

[POP1] T.Popoviciu, Remarques sur la définition fonctionelle d'un polynôme d'une variable réelle, *Mathematica (Cluj)*, **12**(1936) 5-12.

[POP2] T.Popoviciu, Sur les solutions bornées et les solutions mesurables de quelques équations fonctionelles, *Mathematica (Cluj)*, **14**(1938) 47-106.

[RAO] R.Rao, Characterization of probability laws by linear functions, *Sankhya, Ser. A*, **33**(1971) 265-270.

[REI1] L.Reich, J.Schwaiger, Über algebraische Relationen zwischen additiven und multiplikativen Funktionen, *Aequationes Math.*, **27**(1984) 114-134.

[REI2] L.Reich, J.Schwaiger, On polynomials in additive and multiplicative functions, in "Functional Equations: History, Applications and Theory", Publishing Company, Dordrecht, Boston, Lancester, 1984, 127-160.

[RES] S.G.Restrepo, Continuidad de polinomios en espacios vectoriales topologicos, *Revista Columbiane de Math.*, **13**(1979) 271-310.

[RIS] J.Riss, Transformation de Fourier des distributions, *C.R.Acad.Sci.Paris*, **229**(1949) 12-14.

[ROS] R.A.Rosenbaum, S.L.Segal, A functional equation characterizing the sine, *Math. Gaz.*, **44**(1960) 97-105.

[ROT] G.-C.Rota, R.Mullin, On the foundations of combinatorial theory, in "Graph Theory and its Applications", Academic Press, 1970.

[RUD] W.Rudin,"Fourier Analysis on Groups", Interscience, New York, 1962.

[RUK1] A.L.Rukhin, The solution of functional equations of d'Alembert's type for commutative groups, Mimeograph Series 79-23, Dept. of Statistics, Purdue University, 1979.

[RUK2] A.L.Rukhin, The solution of the functional equation of d'Alembert's type for commutative groups, *Internat. J. Math. Math.Sci.*, **5**(1982) No.2.

[SAK] G.N.Sakovič, Functional equations for sums of exponentials, (Russian), *Publ. Math. Debrecen*, **11**(1964) 1-10.

[SAT] R.Sato, A study of functional equations, *Proc. Phys. Math. Soc. Japan*, **(3)10**(1928) 212-222.

[SCH] H.Schmidt, Über das Additionstheorem der Binomialkoeffizienten, *Aequationes Math.*, **7**(1974) 302-306.

[SCW] J.Schwaiger, On the matrix equation $A(t + s) = A(t)A(s)$, Talk given at the 20th ISFE in Oberwolfach, Germany, 1982., (Report of Meeting, *Aequationes Math.*, **24**(1982) 283.)

[SCZ1] L.Schwartz, Théorie génerale des fonctions moyenne - périodiques, *Ann. of Math.*, **(2)48**(1947) 857-929.

[SCZ2] L.Schwartz, Sur une propriété de synthèse spectrale dans les groupes non compact, *C. R. Acad. Sci. Paris*, **227**(1948) 424-426.

[SEG] S.L.Segal, On a sine functional equation, *Amer. Math. Monthly*, **70**(1963) 306-308.

[SEG2] I.E.Segal, The group algebra of a locally compact group, *Trans. Amer. Math. Soc.*, **61**(1947) 69-105.

[STA] P.Stäckel, Sulla equazione funzionale $f(x + y) = \sum_{i=1}^{n} X_i(x)Y_i(y)$, *Atti Accad. Naz. Lincei Rend. Cl. Sci. Fis. Mat. Nat.*, **(5)22**(1913) 392-393.

[STE] C.Stephanos, Sur une catégorie d'équations fonctionelles, *Rend. Circ. Mat. Palermo*, **18**(1904) 360-362.

[STO] J.J.Stone Exponential polynomials on commutative semigroups, Appl. Math. and Stat. Lab. Technical Note No.14, Stanford University, 1960.

[SWE] L.Sweet, On the generalized Cube and Octahedron functional equation, *Aequationes Math.*, **22**(1981) 29-38.

[SWI1] H.Swiatak, On Two Functional Equations Connected with the Equation $\varphi(x+y)+\varphi(x-y) = 2\varphi(x)+2\varphi(y)$, *Aequationes Math.*, 4(1970) 260-261.

155

[SWI2] H.Swiatak, On Two Functional Equations Connected with the Equation $\varphi(x+y) + \varphi(x-y) = 2\varphi(x) + 2\varphi(y)$, *Aequationes Math.*, **5**(1970) 3-9.

[SZÉ1] L.Székelyhidi,"On a Class of Linear Functional Equations", Thesis, University of Debrecen, Debrecen, 1977.

[SZÉ2] L.Székelyhidi, Functional equations on ordered fields, *Publ. Math. Debrecen*, **24(1-2)**(1977) 173-179. (with Z. Daróczy and K. Lajkó).

[SZÉ3] L.Székelyhidi, Remark on a paper of M. A. McKiernan, *Ann. Polon. Math.*, **36**(1979) 245-247.

[SZÉ4] L.Székelyhidi, Almost periodic functions and functional equations, *Acta Sci. Math. Szeged*, **42**(1980) 165-169.

[SZÉ5] L.Székelyhidi, An extension theorem for a functional equation, *Publ. Math. Debrecen*, **28**(1981) 257-279.

[SZÉ6] L.Székelyhidi, Functional equations on Abelian groups, *Acta Math. Acad. Sci. Hung.*, **37**(1981) 235-243.

[SZÉ7] L.Székelyhidi,"Exponential Polynomials and Functional Equations on Topological Groups", University of Debrecen, Debrecen, 1982.

[SZÉ8] L.Székelyhidi, On a class of linear functional equations, *Publ. Math. Debrecen*, **29**(1982) 19-28.

[SZÉ9] L.Székelyhidi, Note on exponential polynomials, *Pacific J. Math.*, **103(2)** (1982) 583-587.

[SZÉ10] L.Székelyhidi, On the zeros of exponential polynomials, *C. R. Math. Rep. Acad. Sci. Canada*, **4(4)**(1982) 189-194.

[SZÉ11] L.Székelyhidi, Almost periodicity and functional equations, *Aequationes Math.*, **26**(1983) 163-175.

[SZÉ12] L.Székelyhidi, Local polynomials and functional equations, *Publ. Math. Debrecen*, **30**(1983) 283-290.

[SZÉ13] L.Székelyhidi, Regularity properties of polynomials on groups, *Acta Math. Acad. Sci. Hung.*, **45(1-2)** (1985) 15-19.

[SZÉ14] L.Székelyhidi, Regularity properties of exponential polynomials on groups, *Acta Math. Acad. Sci. Hung.*, **45(1-2)**(1985) 21-26.

[SZÉ15] L.Székelyhidi, A characterization of trigonometric polynomials, *C. R. Math. Rep. Acad. Sci. Canada*, **7(5)**(1985) 315-320.

[SZÉ16] L.Székelyhidi, Exponential polynomials and differential equations, *Publ. Math. Debrecen*, **32**(1985) 105-109.

[SZÉ17] L.Székelyhidi, The Fourier transform of exponential polynomials, *Publ. Math. Debrecen*, **33(1-2)**(1986) 13-20.

[SZÉ18] L.Székelyhidi, The Fourier transform of mean periodic functions, *Util. Math.*, **29**(1986) 43-48.

[SZÉ19] L.Székelyhidi, On addition theorems, *C. R. Math. Rep. Acad. Sci. Canada*, **9(3)**(1987) 139-141.

[SZÉ20] L.Székelyhidi, Fréchet equation and Hyers's theorem on noncommutative semigroups, *Ann. Polon. Math.*, **48**(1988) 183-189.

[SZÉ21] L.Székelyhidi, On the Levi-Civitá functional equation, *Berichte der Math.-Stat. Sektion der Forschungsgesellschaft Joanneum-Graz*, **301**(1988).

[SZÉ22] L.Székelyhidi, On a linear functional equation, *Aequationes Math.*, **38(2-3)**(1989) 113-122.

[SZÉ23] L.Székelyhidi, Note on polynomial mappings, *Publ. Math. Debrecen*, **36**(1989) 263-265.

[SZÉ24] L.Székelyhidi, Summing power series with exponential polynomial coefficients, *Publ. Math. Debrecen*, (to appear)

[SZÉ25] L.Székelyhidi, On a problem of S.Mazur, *Transactions of AMS*, **316(1)** (1989) 161-164.

[TSC] ,"The Scottish Book", Boston, 1981.

[UNG1] A.Ungar, Generalized hyperbolic functions, *Amer. Math. Monthly*, **89**(1982) 688-691.

[UNG2] A.Ungar, Addition theorems for solutions to linear homogeneous constant coefficient ordinary differential equations, *Aeq. Math.*, **26**(1983) 104-112.

[VIE] L.Vietoris, Zur Kennzeichnung des Sinus und verwandter Funktionen durch Funktionalgleichungen, *Jour. reine und ang. Math.*, **186**(1944) 1-15.

[VIN1] E.Vincze, Über die Verallgemeinerung der trigonometrischen und verwandten Funktionalgleichungen, *Ann. Univ. Sci. Budapest Eötvös Sect. Math.*, **3-4**(1960/61) 389-404.

[VIN2] E.Vincze, Über eine Verallgemeinerung der Pexiderschen Funktionalgleichungen, *Studia Univ. Babeş-Bolyai Math.*, **7**(1962) 103-106.

[VIN3] E.Vincze, Eine allgemeinere Methode in der Theorie der Funktionalgleichungen I., II., III., IV., *Publ. Math. Debrecen*, **9-10**(1962/63) 149-163; 314-323; 191-202; 283-318.

[WIL1] W.H.Wilson, On a certain general class of functional equations, *Bull. Am. Math. Soc.*, **23**(1916) 392-393.

[WIL2] W.H.Wilson, On a certain general class of functional equations, *Am. J. Math.*, **40**(1918) 263-282.

[WIL3] W.H.Wilson, On Certain Related Functional Equations, *Bull. Amer. Math. Soc.*, **26**(1919-20) 300-312.

[WIL4] W.H.Wilson, Two general functional equations, *Bull. Am. Math. Soc.*, **31**(1925) 330-334.

[B120] C.J. Knight, On ablation theorems. Q.Jl. Mech. appl. Math. ...
Canada, 8(30)(1957) 303-311.

[B12?] L. Sarason, Some interior conditions and Cauchy theorems for over-determined systems. Arch. Ration. Mech. 16(1964) 183-190.

[B12?] ... R. Kremer ... in ... Levi-Civita ...

[B12?] ... Lynch, On a linear functional equation. Aequationes Math. ...
19(1989) ...

[B123] L. Schaefli ... Monatsber. ... K.P. Akad. ...
30(1860) 293-295 ...

[B126] L.S. ... Sur une nouvelle série ... partial ...
Leipz. Ber. Math. Classe ...

[B12?] L.E.J. ... OAN, reply in S. Mann, Transactions of ... 11 ...

[150] ... Phil. ...

[156]] ... Laser Math. Transfer ...
600 ...

[1??] A. ... für optimale ... optimale ... für ...
optimale ... Math. Z.(19??) ...
... Math. Ann. ...

[??] A. ... Über die ... ein Lösung ...
... Würll ... Nova Acta ...

[??] ... über ... Differential ... Rend ...

[?1??] ... Math. ...
Jacobi(2) 3 ... III, IV. Pell ... Math. ...
III, 232 ... 332-373.

[?1?] ... über ... Math. ...
Ann. 82 ... 361, 368-393.

[W1?] ... Math. 40(1918) 27-154 ...

... Reine ...
Jber. Sch ... 26(1917) 370-400, 374 ...

[W1?] ... Über ... funktional ... Math. ...
II(19?) 1670-673.